Popular Science

Do-it-yourself Yearbook

Popular Science

Do-it-yourself Yearbook

1984

 New York

 VAN NOSTRAND REINHOLD COMPANY
New York   Cincinnati   Toronto   London   Melbourne

Published by

Popular Science Books
Times Mirror Magazines, Inc.
380 Madison Avenue
New York, NY 10017

Distributed to the trade by

Van Nostrand Reinhold Company
135 West 50th Street
New York, NY 10020

ISSN: 0733-1894

ISBN: 0-442-27489-0

Manufactured in the United States of America

introduction

Welcome to the second annual *Popular Science Do-it-yourself Yearbook*. In it you'll find over 20 chapters on trends and techniques in house construction, remodeling, and repairs. Plus you'll find over 40 woodworking projects for use indoors and out. As well, there are excellent tool-technique chapters.

Chapters for this Yearbook were selected mainly from recent issues of *Popular Science* and *Homeowners How To* (now *The Homeowner*), with two chapters from *Mechanix Illustrated* and a chapter from *Workbench*. Just before press time, the articles were updated and adapted by those who created them, so you can count on their timeliness and authority.

Do-it-yourself writing

We who write in the do-it-yourself field have banded together to form the National Association of Home and Workshop Writers (NAH&WW). Many of the writers in this Yearbook are Association members, who include full-time freelancers, staff editors of do-it-yourself publications, and freelance moonlighters who earn their primary living in other fields. We have noted a resurgence of interest in project articles and a continuing interest in articles on home improvement, maintenance, and repairs. We've also found that readers are interested in doing a wide variety of work. While beginners still need basic instructions, many readers are looking for advanced skills and tips. As a case in point, in the early 1960s, I tried in vain to interest magazine readers in do-it-yourself homebuilding. Now such articles and books are highly popular, and owner-builder schools are springing up across the country.

Of course, painstaking research goes into many of these stories. As a full-time writer myself, I can testify to the challenge of writing concise instructions and conceiving lucid illustrations. For a model presentation, turn to Bernie Price's chapter on building an elaborate set of colonial cabinets, beginning on page 76. Such project articles are the toughest of all to do. First, Bernie had to design the project and collect the tools and materials to build it. Next he built it, stopping at every step to take a photograph. Finally he had to write the story, print the photos, and make detailed sketches for the artist, Carl De Groote. Then he had to write the captions and present an "editorial package" that appealed to editors.

Some Yearbook highlights

Two acclaimed masters of home-shop writing and photography are Ro Capotosto and R.J. (Cris) DeCristoforo.

Beginning on page 132, Ro shows how to build a wrap-around tree bench. Don't miss his tips on miter-cutting the seat boards. Cris DeCristoforo has found so much reader interest in his articles on jigs for power tools that he now provides full-size template plans by mail order. For examples of his many jigs, see pages 104 and 109. Cris' multi-sheet sets of plans measure 30×40 inches. He mails them rolled, first-class for $14.50 postpaid. At this writing, plans are available for Cris' table-saw master jig and a drill-press jig. Other plans are on the drawing board. For the latest catalog, send $1 to DeCristoforo Designs (YB), 27082 Horseshoe Lane, Los Altos Hills, CA 94022.

For advanced woodworking with a touch of trigonometry, R. Joseph Ransil explains how to calculate saw angles for compound miters on page 144. Anyone who has a radial or table saw and does much with it should appreciate Joe's synergistic combination of geometry, math, and woodworking. It's worth filing. Back to basics, Mack Phillips' feature on sharpening shop tools on page 148 is worth saving too. So is Tom Jones' guidance on drilling big holes—page 151. And, for a collection of outdoor seats styled by pro architects, see John Robinson's portfolio beginning on page 135.

Another chapter that stands out in my mind is "Secrets of Drywall Taping" by Mark Lee Due, an expert drywall installer (page 164). Drywall joint taping has always intrigued me because I can't do it well. Mark's chapter showed me what I'd been doing wrong and how to proceed next time with confidence.

More in store

The chapters above are just some of those by freelancers. You'll also find many excellent chapters by the editorial staffs of *Popular Science* magazine and Popular Science Books. Whether your interests and abilities run from basic to advanced, you'll find plenty in this *Popular Science Do-it-yourself Yearbook*.

RICHARD DAY
President, National Association of
Home and Workshop Writers and
Consulting Editor for Home and Shop,
Popular Science

the authors

Below is an alphabetical sampling of the authors for this issue. They include freelancers as well as members of the *Popular Science* New York and field staff.

Erik H. Arctander, a former senior editor with *Popular Science,* works as a deck officer on merchant ships. In port, he busies himself with the renovation of his Victorian landmark home and writes freelance. He is the author of books on motorcycles and boating.

Paul Bolon took a humanities and science degree from M.I.T. before working several years as a cabinetmaker and carpenter. Later, in New York as an associate editor for *Popular Science,* he wrote and edited articles on science, consumer products, and do-it-yourself projects. He's now writing full-time, also renovating his Chicago condo and indulging in its basement workshop.

Rosario Capotosto, a master woodworker and photographer, is one of the most sought-after writer-craftsmen by editors in the do-it-yourself field. Known best for his woodworking, Ro also works in metals and plastics, and he writes on home improvements and maintenance as well. His books include *The Complete Book of Woodworking, Working with Plastics, Capotosto's Woodworking Techniques and Projects,* and *Capotosto's Woodworking Wisdom.*

Frank H. Day took an engineering degree from Stanford University and then worked 44 years in mining and metallurgy in Mexico and the U.S., winding up as manager of a large metallurgical plant in Montana. Since retiring, Frank has published shop hints and do-it-yourself articles on woodworking and photography.

Richard Day, contributing editor for home and shop at *Popular Science,* was the owner of an auto-repair shop before he turned to writing. Considered an expert in many subjects beyond auto repair, such as masonry, plumbing, housebuilding, shop work, and tools, Day is a prolific writer of magazine do-it-yourself articles and the author of over a dozen related books, including *How to Build Patios and Decks* and *How to Service and Repair Your Own Car.* He is also president of the National Association of Home and Workshop Writers.

R. J. De Cristoforo has long been one of the leading woodworking writers. He is widely recognized as a master of tools, a shop-work genius, a highly accomplished photographer, and a skilled draftsman. Besides serving as consulting editor for tools and techniques for *Popular Science,* he is the author of countless articles and over two dozen books, including *How to Build Your Own Furniture; Build Your Own Wood Toys, Gifts and Furniture;* and *De Cristoforo's Housebuilding Illustrated.*

Carl De Groote, author of the chapter "Bi-level Outdoor Living," is a prolific architect/illustrator for the major do-it-yourself publications. Renowned for his knowledge of construction and for his attention to detail, Carl is often asked to produce complex drawings from mere snapshots. His drawings illustrate nearly half of the chapters in this Yearbook. They are easily recognizable by their style and are usually signed CDG.

Patrick J. Galvin is former editor and publisher of *Kitchen & Bath Business* magazine and has written more than 5,000 stories as a freelance writer/photographer. His books for homeowners include *Remodeling Your Bathroom, Finishing Off, Book of Successful Kitchens,* and *Successful Space-saving at Home.* His other books include a professional volume, *Kitchen Planning Guide for Builders, Designers and Architects.*

A. J. Hand began preparing for a career as a writer-photographer at the age of six by assisting his writer-father, the late Jackson Hand, with photo setups. A. J. later served on the *Popular Science* staff, winding up in 1975 as the magazine's home workshop editor before he began freelancing full-time. His photos have appeared on many magazine covers, and his articles appear in magazines and in his syndicated newspaper column, "Hand Around the House." He is the author of *Home Energy How-to.*

W. David Houser, art director for *Popular Science,* has written up several of his home projects for the magazine. At home in upstate New York, Dave is usually working on some improvement or woodworking project. For this, Dave does his own designing. He also restores old MGs. One of his favorite models, a 1953 MG-TD, is partially visible in Dave's shop, shown on page 97.

Cathy Howard is a freelance writer specializing in home improvement, gardening, travel, medical and other consumer-interest features. Her career includes work as an advertising and public relations specialist for several major corporations as well as regular writing assignments for *Better Homes and Gardens, Woman's Day, The Homeowner,* and other national publications.

Herb Hughes is a technical writer for Intergraph Corporation and has worked in home-design and construction fields since the mid-1960s. Herb's many articles have appeared in the major do-it-yourself magazines, including *The Homeowner.* He has published two books: *More Living Space* and *Home Remodeling Design and Plans.*

John H. Ingersoll has been an editor and writer for over 30 years, specializing in housebuilding, remodeling, energy, architecture, and science. Formerly senior editor of *House Beautiful,* John has written nearly 1,000 articles for many magazines including *The Homeowner* and *Popular Science.* He is the author of four books, including *How to Buy a House.* John is currently president of the American Society of Journalists and Authors.

Thomas H. Jones is a full-time writer specializing in furniture making, home improvement, and woodworking techniques. Before taking up writing full time in 1970, Tom was an aerospace engineer. He has sold hundreds of articles on do-it-yourself subjects and is the author of three books, including *How to Build Greenhouses, Garden Shelters and Sheds.*

Charles A. Lane is an underground construction consultant based in St. Paul and a partner in Terra Builders Inc., a development firm now building a 100-unit underground housing project in Denver. He is a former assistant director of the University of Minnesota's Underground Space Center. Charles is also a recreational scuba diver and admits that his interest in things under water and under ground may relate to his fear of heights.

Lloyd Lemons, Jr., author of "How to Wire a Post Lamp," has been a journeyman electrician, a machinist, and an auto mechanic. He also enjoys woodworking and doing his own home improvements. He is now the owner of an advertising agency and a partner in a publishing firm.

E. F. (Al) Lindsley is the *Popular Science* senior editor for engineering. He's also been a steeplejack, a semi driver, a carpenter, a meteorologist, a test pilot, an industrial engineer, and an editor with *Scientific American.* For *PS* and *The Homeowner* he covers motors, engines, vehicles, house repairs, and woodworking. Al is the author of the long-time standard *Engine Installation Manual* for the Internal Combustion Institute as well as *Electric and Gas Welding* and *Metalworking in the Home Shop.*

Charles A. Miller, associate editor at *Popular Science,* came to *PS* in 1982 from *Mechanix Illustrated,* where he handled virtually all of *MI's* science, energy, and technology stories. At *PS* he handles how-to, science, and consumer stories. Chuck has designed and built a variety of well-received stories on decks, greenhouses, and room additions—some of which appear here.

Jeff Milstein is an award-winning architect. Educated at Berkeley, he now lives and works in Woodstock, New York. He has been designing home projects and writing on passive solar topics for *Popular Science* since the early 1970s, and his designs have been published internationally. He is coauthor of *Designing Houses: A Guide To Designing Your Own House.*

Kenn Oberrecht began building furnishings and writing about them before finishing his M.A. in journalism at the University of Alaska. He has since filled several apartments and two houses with his woodworking projects. Kenn calls coastal Oregon home but also lives half the year in a log cabin on Chichagof Island, Alaska. He's a full-time writer and photojournalist, specializing in outdoor-recreation and do-it-yourself topics, and the author of seven books, including *Plywood Projects Illustrated.*

Evan Powell, Southeast editor for *Popular Science* and *The Homeowner,* specializes in home repairs, equipment, remodeling, and energy systems. Earlier Evan was Technical Service Director for Sears, Roebuck and Co. He is a product consultant for General Electric and helps produce their "Quick-Fix" program. He also produces home equipment features for Multimedia TV and "Road Test" for Pulitzer TV. He is the coauthor of *The Complete Guide to Home Appliance Repair* (with Robert P. Stevenson) and *The Popular Science Book of Home Heating and Cooling* (with Ernest V. Heyn).

Bernard L. Price, a former mechanical engineer and head technical writer for Edmund Scientific Corporation, has worked 12 years in automotive repair and 11 years in cabinetmaking and general construction. A full-time freelance writer now, Bernie is a principal contributor to major do-it-yourself magazines and book collections. His specialties include tools, machines, plumbing, wiring, cabinetmaking, remodeling, and outdoor structures.

John Robinson is a veteran writer, editor, and photographer in the architectural and home fields. He often generates do-it-yourself magazine articles as by-products of a home problem he's been forced to solve, simply photographing his steps in case somebody else might need guidance. He is the author of books on travel and *Highways and Our Environment.*

Benjamin T. (Buck) Rogers is an engineer in private practice with special expertise in solar design. He was a founding member of the Los Alamos National Laboratory (LANL) Solar Group and consults for LANL in the analysis of passive solar systems. Prior to entering private practice he was for nearly three decades a member of the University of California's Los Alamos Laboratory. He has taught energy-efficient architecture for Arizona State University and a field architecture course for Pratt Institute.

Daniel J. Ruby, an associate editor at *Popular Science,* writes and edits articles on technology, energy, physical sciences, house design, and home do-it-yourself projects. In recent years, he has honed his construction and home-improvement skills by converting a raw warehouse space into a living loft. His newest interest is his home computer, upon which he coauthored (with Ernest Heyn) the *Popular Science Book of Home Alternate Energy Projects.*

John Sill is vice president and editorial director of Popular Science Books. As his Yearbook chapter on the Queen Anne Table attests, John's spare time interest turns to woodworking, a prime love of his life since he picked up a spokeshave in a fifth-grade manual training class. He particularly enjoys building replicas of fine furniture and designing and building modern furniture for articles appearing in magazines such as *Popular Science* and *Workbench.*

V. Elaine Smay, a senior editor at *Popular Science,* writes and edits stories on diverse subjects including house design, home heating and cooling systems, camping equipment, and tools. She's become the magazine's expert on earth-sheltered housing, a feature in this Yearbook. She began running in 1968, when running wasn't cool, and has since competed in marathons. At home she enjoys tending her hydroponic garden.

Neil Soderstrom is a senior editor with Popular Science Books. His chapters in this Yearbook on a sawbuck and a woodpile shelter/ fence reflect his enthusiasm for weekend lumberjacking. He also writes on a broad range of home-improvement subjects. His books include *Chainsaw Savvy, Heating Your Home with Wood,* and (with E. D. Fales) *How to Drive to Prevent Accidents.*

Richard Stepler, group editor for consumer information at *Popular Science,* has written on a variety of home-improvement subjects, ranging from house design and construction to decks, roofing materials, lighting, and custom built-ins. He and his wife and young son live in a 100-year-old cast-iron-front loft building in New York's Soho, which has been a source of home remodeling stories for *PS.*

Peter and Susanne Stevenson are the husband-and-wife team behind Stevenson Projects, Inc., a California company that supplies do-it-yourself plans for scores of projects ranging from furniture to sailboats. Many of their projects, which combine beauty and utility, have been featured in national magazines. Pete's the designer, builder, writer. Susie's the business manager and publications director.

contents

V.
Woodworking and Tool Techniques 142

VI.
Plumbing, Wiring, Taping 156

VII.
Repairs and Maintenance 168

VIII.
The Latest in Tools 172

double-shell houses

This is a controversial house?" I thought, with something close to incredulity, as I entered an asphalt drive in Middletown, Rhode Island, a Newport suburb, and faced a gray-shingled Cape Cod cottage, a pleasant but ordinary rendering of a New England stereotype.

Once inside, though, it became apparent that this was no ordinary Cape Cod. The afternoon sun streamed in through skylights in a mostly glass solarium, spotlighting hanging plants and pots of white chrysanthemums. From inside it looked like a passive-solar house.

But it's not an ordinary passive-solar house, either. I got my first confirmation of its true thermal nature when I looked out a north window. Beyond the glass, a few inches away, was a second window. Wood slats, a fraction of an inch apart, bridged the gap between on all four sides. Looking closely between the slats I could see that the wall was hollow. And that's where the controversy begins.

The house, which belongs to Robert and Elizabeth Mastin, was one of the first double-shell houses built. The north wall is one part of a complete air plenum that envelops the house on four of its six sides. It includes the solarium, the attic, the space in the north wall, and a crawl space under the lower level. According to the original theory of double-shell houses, heated air from the solarium is supposed to flow by natural gravity convection around that plenum, distributing the heat evenly to the house and storing the excess in the crawl space and the earth below for use at night.

No sooner had one of these houses been built and publicized than the critics attacked. They started by disputing the basic theory, then went on to question performance claims, cost effectiveness, and even the safety of a house with a plenum through which fire might quickly spread. Just as vociferous were the proponents: owners who lived in the houses, architects who designed them, and others who just liked the idea of an energy-efficient house with no mechanical systems, no massive masonry walls, no jugs of water, and plenty of windows. About the only thing all agreed on was the need to monitor some of these houses and see: (1) if they work, and (2) how they work.

Now such studies have been done and the results analyzed. One of the most comprehensive used the Mastin house as its guinea pig. The conclusion: The house works—but for the wrong reasons.

Mastin house, a classic double-shell begun in 1978, was recently modified to improve performance. Roof glazing (photo above) was removed and replaced with much smaller area of operable skylights (photos left and right). Air plenum for convective air flow was blocked, and insulated ducts and a thermostatically controlled fan were installed to draw hot air from the top of the solarium and deliver it to crawl space below the lower level, to be stored for use at night, as shown in drawings, next page. Mastin (shown with author in photo left) also removed insulation from attic roof and beefed up ceiling insulation to R-60.

With that evidence in hand, Mastin, who has his own design firm (Natural Energy Design, Inc., 1355 Green End Avenue, Middletown, RI 02840), and a few others are modifying the basic double-shell concept in ways that may make it work even better.

The Brookhaven tests

In the winter of 1980, a team from the Department of Energy and the Environment of Brookhaven National Laboratory delivered a carload of instruments to the Mastin house. They installed thermistors to measure temperatures in nine locations, three recording hygrometers to measure relative humidity, and two recording pyranometers to measure solar insolation. To measure the exact amount of auxiliary heat used they installed a 1500-watt electric heater on each floor, each controlled by a thermostat set at 65 degrees F. (The Mastins were instructed to keep their fireplace and wood stove cold.) Monitoring took place in parts of January, February, and March. More tests were run in July. The following winter, thermocouples were installed to monitor the temperatures below the slab. "We collected data off and on for almost a year," says Mastin. "My wife got a little tired of all those instruments and wires hanging around."

It took Brookhaven months to analyze all the data. But finally the report was published. To begin with, it confirmed that the house needs very little auxiliary heat. "It used less than one-quarter the purchased energy of a recently built conventional house of comparable space," the report states.

But the tests didn't do much to confirm the theory. Stratification was an obvious problem. For a typical week in January, on sunny days the temperature in the attic always rose to 90 degrees, topping out at 106. In the crawl space, however, temperatures remained between 45 and 55 for the entire period. Temperatures on the three living levels reflected a corresponding stratification. "Auxiliary heat is needed on the lower level even when the upper level is overheated," the Brookhaven report notes. During

a typical two-day period in January, the heater on the top floor was on 5 percent of the time while the one on the lower level was on 82 percent. Clearly, gravity convection wasn't distributing heat evenly through the house.

Next, the Brookhaven team attempted to measure the convective flow directly. Actually, they instructed Mastin, who holds an engineering degree from the Naval Academy at Annapolis, Maryland, and he did the legwork. "They gave me a hand-held instrument called an ionized corona probe," he reports. "I did measurements at various times and all over the envelope."

What did he see? "In the heat-gain mode [a sunny day] I definitely saw air flow in the direction the theory predicts," he reports, "but it was fairly slow. At night, there wasn't any definite, consistent flow pattern. Most of the time it was just confused. But it was never static." Theory said it should reverse at night and flow in the direction opposite to daytime flow.

After analyzing the data Mastin turned in, the Brookhaven team concluded that the directions were too confused to be meaningful and the velocities definitely low. They pointed out, however, that since the volume of air is so large, a significant amount of heat could be moved even with the low velocities.

So Mastin devised a simple test to see how much the convective flow contributed to heating the house. He took building paper and a staple gun and covered the opening between the crawl space and the solarium, blocking the loop. The result: The blockage caused no performance degradation. In fact, the auxiliary heat needed was "lower by 5 to 20 percent (depending on insolation) with the loop blocked," Brookhaven reports.

Other tests indicated that there was little effective heat storage in the crawl space and the earth below. One way they checked this was by comparing the auxiliary heat requirements for two nights with similar outside temperatures. The first night followed a sunny day; the second, a cloudy day. If, indeed, solar heat is stored in the daytime and released to the loop at night, less auxiliary heat should have been required after the sunny day. In fact, slightly

3

less was required after the cloudy day. Still, the warmth from the earth itself did keep the solarium above freezing at all times.

Having lived in the house for a year, Mastin had surmised that it had too much glazing in the solarium roof. To test that, he installed fiberglass batts over the upper half of the double row of windows, blocking both solar gain and heat loss. With the insulation in place, "the auxiliary heat required was reduced by 25 to 50 percent," Brookhaven's analysis revealed.

Limited tests done in July indicated that the house stays cool in summer as well as it stays warm in winter. For example, one July day when the temperature outside reached 87 degrees F and the solarium rose to equal that, the living areas of the house ranged from 71 to 74, without auxiliary cooling. Certainly the outer shell, which shades the living area from the sun, contributed to this. But the cooling tubes, a standard feature of double-shell houses in some climates, didn't seem to help much. In cooling mode, hot air is supposed to rise from the solarium and exit through attic vents, pulling fresh air that has been cooled and dehumidified by earth contact into the envelope through the cooling tubes. During part of the Mastin tests, air flow was in the wrong direction: It flowed from the house, through the cooling tubes, and out what was supposed to be the inlet.

(Actually, the Mastin house doesn't present much of a cooling challenge. It's about a mile from the Atlantic, and prevailing southwest winds bathe it in cool ocean air on most summer afternoons.)

After analyzing all the data collected, the Brookhaven team reached this conclusion: "The low energy needs of the Mastin house are attributable mainly to the excellent insulative value of its double shell." If cost were a simple overriding consideration, the report continues, "it would be difficult to rationalize the double envelope's use over the super-insulated house."

Christopher Shipp house (see drawing, right).

Super-insulated houses are exactly what the name implies. They are constructed to include an enormous amount of insulation. They conserve heat so well that internal heat gain—from people and appliances—is nearly enough to carry the house, even in a severely cold climate. They rely on solar heat very little, and hence have small windows, even on the south, since windows can be a large source of heat loss.

Hybridization

But, Mastin decided, why not have the best of both worlds? His current designs blend many aspects of a super-insulated house with some vestiges of the double-shell design. He calls the result a Hybrid Geotempered Envelope (HGE).

On a lovely wooded lot south of Boston, builder Robert Green completed an HGE house for Robert and Nina Heyd (see photo). Gary and Kathy Brennan and their five children recently moved into another near Little Compton, Rhode Island. Both have solariums, but there's no air plenum forming a loop around these houses. Instead, a

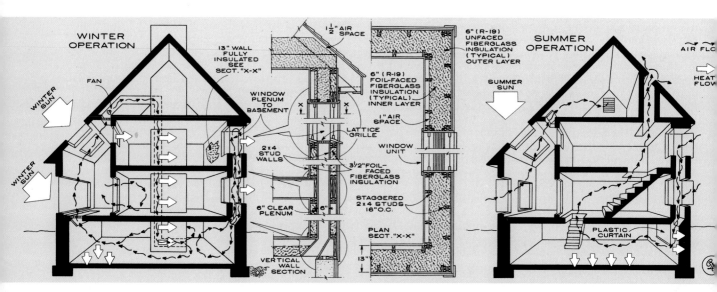

Mastin's Hybrid Geotempered Envelope design has super-insulated walls on north, east, and west: double-stud walls with 12-inch (R-38) insulation. Attic insulation is R-60; foundation, R-10. Double windows have plenums that open into basement. On a sunny winter day (left), 800-cfm fan draws hot air from top of solarium through insulated duct in attic and down a duct in masonry chimney. There it loses heat to the mass. Duct opens into basement, where warm air loses more heat to the mass there and to the earth below. Air is then drawn up through grates in solarium floor, and cycle repeats. In window plenums, air near outer glass cools and falls to basement, displacing warmer air, which rises into plenums. On winter nights, solarium is kept warm in similar way, and masonry chimney releases heat to house. In summer (right) attic vents draw solarium heat through opened trap door. Attic fan draws air through open outer windows, down plenums to basement, up stairwell, out attic.

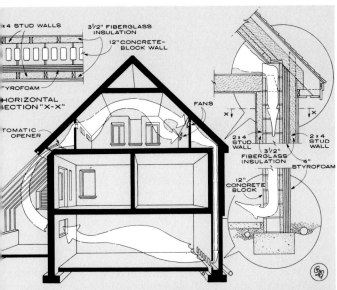

Air flow around loop in house designed by Christopher Shipp is fan-forced. On winter days, as sunspace warms, insulated doors at top open. Fans near north wall pull warm air across upper story and down plenum in super-insulated north wall (R-40). Plenum is formed by voids in concrete blocks (for fire resistance and heat storage). (Other walls are R-30, roof is R-45, foundation is R-20.) Air exits from north wall to lower level, where more heat is stored, then returns to sunspace, and cycle repeats. At night, insulated doors at top of sunspace close. In summer they are also closed. Then warm air from house is pulled down by fans through north wall and dumped in lower level, losing heat to mass. It returns up stairwell. Root cellar (not shown) doubles as heat exchanger.

"They only spent $60 on electricity to heat the house last winter, so spending $200 on fans seems needless," Shipp reports.

How do these hybrid designs compare in cost, performance, and livability with super-insulated houses?

"The best-performing houses anywhere are the super-insulated houses," Mastin admits. "They are also the most cost-effective to build. But their obvious shortcoming is lack of light, views, and natural ventilation."

If you count the solarium as an extra expense (about the same per square foot as the rest of the house), Mastin's HGE-type houses cost a good bit more to build than superinsulated houses. "But if you compare the cost with that of a super-insulated house with a sunspace," Mastin says, "the only cost increase is for the double windows and their plenums—about $200 per window. And you don't have the cost or nuisance of window insulation."

The auxiliary heat requirements for super-insulated houses generally come out to be even less than what the Brookhaven tests showed Mastin's house to need. Preliminary monitoring of one of Mastin's HGE houses, in South-

Ample window area on all sides of house is one advantage of HGE houses. This is southwest corner of the Heyd house.

thermostatically controlled fan takes hot air from the top of the solarium and delivers it through ducts to the basement.

The attic, foundation, and the north, east, and west walls of these houses are super-insulated (see diagram). The north, east, and west windows, however, look much like the north windows in Mastin's own house: two sets of double-glazed units. There is an air plenum at each window, but it's only as wide as the window and ends at the top (which tends to comfort fire inspectors). It does, however, run all the way down to the basement. "The true effectiveness of the double-shell design is in reducing heat loss through glass," Mastin says. "When the air between the windows is colder than the air in the basement, it's going to drop and displace the warmer air there, which will rise.

"The plenum also reduces infiltration," he points out. "Yet when you need ventilation, you can open an interior window and get it, without a blast of cold air." This mitigates the need for an air-to-air heat exchanger which super-insulated houses often require. And since the windows aren't horrible heat losers, HGE houses can have a good many, even on the north side.

Mastin is not alone in modifying the double-shell concept. In fact, a Pennsylvania designer, Christopher Shipp (R.D. 1, Box 331D, West Grove, PA 19390), has come up with an almost identical design, including both super-insulated walls and double windows with air plenums connected to the crawl space or basement. Shipp is also doing variations of the double-shell house that are closer to the original concept, but using fans to force the air flow (see diagram). At least, he intends that these houses use fans. So far the owners of the house in our photo haven't installed the fans.

ington, Connecticut, suggests that its heat needs are in the same ballpark, though somewhat more than the most efficient super-insulated houses. More such data will be forthcoming. Mastin spent the summer of 1982 modifying his own house to make it more like the hybrid design (see opening page). Brookhaven brought back their instruments and remonitored the house in 1983.

Just what type of house is the most livable is a highly personal matter. "I didn't want a strictly solar house because I didn't want to cut down all my trees," says Robert Heyd, gesturing toward a stand of stately oaks to the south of his house. "And I didn't want a super-insulated house without much window space."

The Brennans are delighted with the choice they made. "I checked out everything," says Kathy. "I visited some passive-solar houses with whole south-facing glass walls. They turned into an oven during the day, then cooled off just as fast at night. I checked out others that had a basement full of rocks, and some with water drums in the living room—that really turned me off. With all the other designs I read about, it seemed we had to give up something. This was the only one that had everything"—*V. Elaine Smay.*

For a free color brochure on Hybrid Geotempered Envelope (HGE) homes, write Natural Energy Design, Dept. PSB, 1355 Green End Avenue, Middletown, RI 02840. Or for $5.75 you can have a detailed information package and design portfolio, including several house plans.

basics for underground homes

The owner-builders of an earth-sheltered house in the Midwest intended to save money on waterproofing by mimicking a respected system but simplifying the application. Result: The roof leaked in twenty-eight places.

A second earth-sheltered house had an off-the-wall problem—literally. Clay soil beneath the footings expanded when saturated with rain. That pushed up the footings, which pushed up structural pipe columns. And they raised the roof 4 inches off the walls.

The roof of another owner-built earth-sheltered house collapsed when the soil was being pushed onto it. The reason: Two of the prestressed-concrete planks had been put on upside down. In that position they were too weak to take the weight.

These may sound like scare stories, but they're not intended to scare. Indeed, for every earth-sheltered-house owner I know who's had serious problems, I've met dozens who are delighted with their houses—and amazed by their low fuel bills.

But those stories are intended to illustrate that an earth-sheltered house demands special considerations in design and special care in construction. If you're thinking of going underground, your first step should be to learn all you can about it. Here are the fundamentals. The sources listed at the end will help you continue your education.

The building site

Successful earth-sheltered houses usually embody good passive-solar design. In a climate that needs heat in winter, the house should be designed to capture solar heat. The obvious way to do this is to orient most of the windows to the south, plus or minus 10 to 15 degrees. North glass will be a serious energy drain in a cold (or even moderate) climate.

When choosing a building site (or if you already have one), you should evaluate its solar potential. An architect with some solar savvy can design around a less-than-ideal lot, but it will be more difficult.

You'll also need to know about the soil and ground water at the building site. You learn this by hiring a soil engineer to do a thorough soil test. He'll extract one or more core samples, then analyze them in the lab. The cost: $500 to $800.

A soil test should tell you:

Type of soil. Some types are better than others for building earth-sheltered structures. A well-drained sand or gravel with a variety of particle sizes, well mixed, is ideal. Some clays expand and contract with changing moisture content. That can be a problem, as the couple whose roof rose 4 inches discovered. A sandy loam is spongy and compresses easily. Rock outcroppings can make excavation very difficult.

Percentage of swelling. If expansive clays are present, the structural engineer needs to know the amount of swelling so he can design for it.

Soil bearing pressure. This tells how much weight the soil can safely support and measures its strength and compaction. It is essential in designing the footings.

Lateral soil pressure. This is a horizontal force that pushes against walls. The deeper the structure, the greater the lateral pressure.

Water table and perched tables. The water table (or aquifer) is the depth of the main ground water below the site. Perched water tables are pockets of water located (perched) above it. When earth-sheltered buildings have water problems, perched water tables are usually to blame. Perhaps the best rule of thumb about ground water is to *never* build in the water table. If it is high, build the house several feet above it and bring in earth to pile on the structure.

Knowing about the soil and ground water of the building site will make the house design considerably easier and will probably save money in construction costs. It will greatly reduce the potential for structural settling and leakage.

Climate

One common misconception about underground construction is that soil is a good insulator. In truth, it is not. The R-value (resistance to heat flow) of relatively dry earth (less than 15 percent moisture) ranges from R-1 to R-5 per *foot*. And it decreases rapidly as soil moisture increases. Extruded-polystyrene insulation board, by comparison, rates an R-5 per *inch*.

Then why bury your house? Because the soil is a terrific moderator of temperature. It does not respond to diurnal temperature variations and only gradually and moderately to seasonal changes. Result: Surrounding the house with soil is similar in effect to building it in a relatively benign climate. Furthermore, the soil's slow response to air-temperature changes causes a thermal lag: Though air temperature may be hottest in July or August, ground temperature probably won't reach its peak until fall. And it will be at its coldest in early spring, not in midwinter.

Another popular misconception is that the temperature underground is a constant 55 degrees F—year-round, at any depth, all over the world. Again, not true.

Ground temperature varies with depth and climate. In Minneapolis, you do find a constant temperature of 55 degrees—but not until you dig down 25 feet. In Phoenix, you'll also find a constant temperature at about 25 feet. But it's 75 degrees, not 55. And nobody builds a house *that* deep underground. Under 2 feet of earth cover (where a roof might be) the yearly temperature in Minneapolis ranges from about 20 degrees to 68 degrees (not bad, when air temperature may

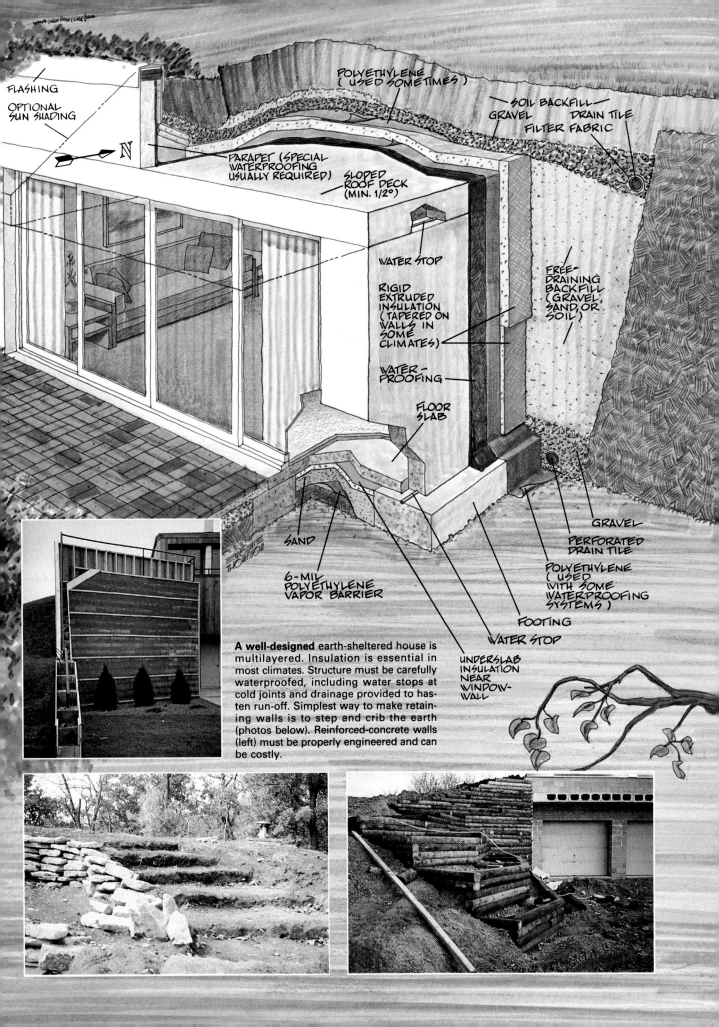

FLASHING

OPTIONAL SUN SHADING

N

POLYETHYLENE (USED SOMETIMES)

SOIL BACKFILL
GRAVEL
DRAIN TILE
FILTER FABRIC

PARAPET (SPECIAL WATERPROOFING USUALLY REQUIRED)

SLOPED ROOF DECK (MIN. 1/2°)

WATER STOP

RIGID EXTRUDED INSULATION (TAPERED ON WALLS IN SOME CLIMATES)

FREE-DRAINING BACKFILL (GRAVEL, SAND, OR SOIL)

WATER-PROOFING

FLOOR SLAB

SAND

6-MIL POLYETHYLENE VAPOR BARRIER

GRAVEL

PERFORATED DRAIN TILE

POLYETHYLENE (USED WITH SOME WATERPROOFING SYSTEMS)

FOOTING

WATER STOP

UNDERSLAB INSULATION NEAR WINDOW-WALL

A well-designed earth-sheltered house is multilayered. Insulation is essential in most climates. Structure must be carefully waterproofed, including water stops at cold joints and drainage provided to hasten run-off. Simplest way to make retaining walls is to step and crib the earth (photos below). Reinforced-concrete walls (left) must be properly engineered and can be costly.

range from −30 to 100 degrees). At 6 feet below the surface, the temperature swings from 36 to 60 degrees. And the swing is from 41 to 54 at 10 feet.

In Phoenix at 2 feet underground, the yearly temperature ranges from 54 to 86 degrees. At 6 feet the range is about 60 to 80 degrees, and at 10 feet it's 68 to 72.

These two extremes in climate require different earth-cover and insulating techniques. In Minnesota, the temperature of the surrounding earth most of the year is colder than the desired room temperature. Thus the house's roof and walls must be wrapped in insulation or the ground will relentlessly bleed heat from them.

In Phoenix, where cooling is the primary energy need, the ground temperature under 2 feet of earth cover is too warm much of the year to pull heat from the house. Insulation can be added to reduce summer heat gain from the soil, but a better solution is to increase the earth cover so that for most of the year the ground temperature is lower than the desired room temperature.

In a moderate climate, where ground temperature is close to the desired room temperature, insulation is often omitted from the lower walls.

Structural systems

Most earth-sheltered houses are made of various types of reinforced concrete, but some use treated wood and some use steel roof decks with concrete poured over them. Here are some points to ponder before you decide on a structural system:

● How much earth cover is best for your climate? In a hot climate where several feet of earth cover is desirable, a wood-beam roof is probably not practical. Under 2 feet of earth cover, however, it can work well.

Under heavy earth cover, arches and domes may be more efficient than flat roof systems because these shapes are inherently strong in compression and thus the shells can be thinner. These shapes can become prohibitively expensive, however, if custom form work must be built. One contractor in Illinois spent more than $100,000 on form work for arched earth-sheltered houses. He plans, of course, to spread the cost over many houses. If ready-made forms are available, domes and arches can be cost-effective.

● What is the lateral pressure exerted on the walls by the earth? (Remember, you learn this from the soil test.) Soils that swell significantly and push against the walls would probably rule out a wood structure. Even concrete

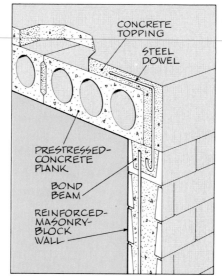
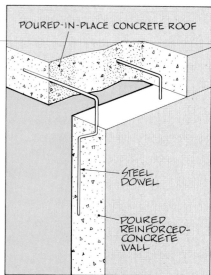
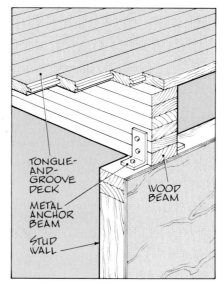
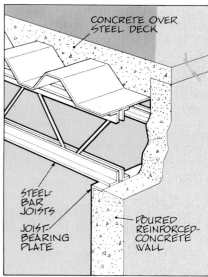

Structural systems differ in strength. Assuming 2 feet of earth cover, here is a rough guide to economical maximum clear spans (between bearing members) of typical roof structures. Top left: pre-stressed-concrete roof planks (10 to 12 inches thick)—22 to 24 feet. Top right: mono-pour or poured-in-place reinforced-concrete roof and walls (8 to 12 inches thick)—24 to 26 feet. Wood beams (4 × 12 laminated or 8 × 10 rough-cut on four-foot centers) with t&g treated decking (2 × 6s)—12 to 14 feet. Above right: steel-bar joists (12 to 18 inches deep on 2-foot centers) and steel deck with concrete topping—16 to 18 feet. Not shown: post-tensioned concrete roof (8 to 10 inches thick)—30 foot wall structures shown can be used with different roofs; however, they must be engineered to handle roof load and lateral soil pressure.

block may not be practical in some cases.

● What is available locally (materials and competent contractors to work with them), and what are the relative costs? Prestressed-concrete-plank manufacturers should be 70 miles or less from your site to avoid high transportation costs. If you're considering a wood structure, be sure what's available is pressure treated with

chromated copper arsenate (CCA) salts, not with creosote or pentachlorophenol. Look for a contractor who puts in All Weather Wood Foundation (AWWF) systems.

● Will the structural system permit the floor plan you want or will bearing members interfere? The captions on the preceding page give a rough guide to maximum economical spans under 2 feet of earth.

Waterproofing

Unfortunately, there is no best water-proofing system for all uses. Waterproofing is a fairly new science with little long-term proof (50+ years) of the products' reliability. Here are the types (and approximate applied costs) of waterproofing materials that, in my experience, have proved to be most reliable: rubberized-bitumen roll goods (90¢–$1.10 per sq. ft.); butyl-rubber sheets ($1.80–$2.60 per sq. ft.); ethylene propylene diene monomer (EPDM) sheets ($1.80–$2.50 per sq. ft.); polymer-mixed bentonite clay products (85¢–$1.10 per sq. ft.).

Bitumens, butyls, and EPDMs should be applied only in roll or sheet form. Bentonite systems work best when sprayed or troweled on. Results have not always been satisfactory when bentonite is put on in paper-encased panels, or when it's poured on dry, then soaked with water (that's what the couple did whose roof leaked in twenty-eight places). The most successful waterproofing systems combine two or more of the above products. Of course, proper backfilling and drainage are essential.

You can find local manufacturers and distributors of waterproofing systems in the Yellow Pages. You'll want to read all the companies' brochures. It may also be worthwhile to take your preliminary house plans to several firms. Even if they charge a consulting fee (about $30 an hour), it will probably be a good investment.

When consulting a waterproofing firm, ask these questions:
● Does the material conform with the structure of your house and anticipated structural movements? For example, liquid-applied elastomers rely on tight adherence to the substrate. Form-release oils, mineral precipitate on concrete, and other factors often prevent adherence, and, as the house settles, leaks develop.
● Is the waterproofing resistant to the chemical nature of your soil, its moisture content, and predicted hydrostatic pressures? Bentonite clays are least effective in soils with a high salt content. An acidic soil plays havoc with modified bitumens. Indeed, highly acidic soils are pretty tough on anything. Probably the best to stand up against them are the synthetic-rubber sheets such as EPDM and butyl. Some manufacturers, including all who make modified bitumens, won't recommend their products for use under standing water. That should preclude their use on underground roofs unless the drainage is excellent or the climate very dry.

● Will the material be effective in the thermal environment underground? Asphalt and pitch products crack and check in cold weather. On a conventional roof, they soften in the sun and the cracks "heal." Underground, it never gets warm enough.
● Are the application requirements compatible with your construction schedule? Some require immediate backfill; others require a 24- to 72-hour cure time.
● What is the warranty? Find out who would pay how much for what if trouble should develop. For example, if there is a leak, who pays to dig away the earth to find it? Who pays to repair it? Who pays to have the earth cover replaced? Who pays for any damage the leakage may have caused? Generally, such water damage would not be covered by a homeowner's insurance policy.

When selecting waterproofing, look beyond the cost. The cheapest product is not always the worst choice, nor is the most expensive always best.

Finally, when selecting your contractor be sure he is certified by the particular product manufacturer to apply that product.

Other considerations

If your earth-sheltered house is to be made of concrete, the insulation should be on the *outside* (see painting) so that you preserve the mass of the concrete for thermal storage. Trouble is, that puts the insulation in contact with soil—and moisture. Consequently, I recommend that only extruded-polystyrene board (such as Dow Chemical's Styrofoam and U.S. Gypsum Co./Condec Corp.'s Foamular, or Minnesota Diversified Products Certifoam) be used. Other kinds of rigid insulation—expanded polystyrene and urethane-based boards—are often used underground, but they tend to absorb moisture, which reduces their R-value. I've dug up a number of houses and have *always* found these types wet and soggy.

A properly built earth-sheltered house will have a very low-air-infiltration rate. That's an advantage for your energy bills, but it could let humidity and even hazardous pollutants build up inside. The house should be designed for natural ventilation when outside weather permits. It should also have a backup fan-driven system, engineered to provide at least one-half air change per hour. I recommend that air-to-air heat exchangers be used with the fan system so heat isn't lost along with the stale air. Bathrooms,

range hoods, and other sources of fumes and odors should be exhausted outside. Each combustion appliance should have its own source of combustion air ducted from outside.

Finally, you should understand that designing and engineering an earth-sheltered house is not a job for the weekend handyman. Even professional engineers and architects may not have the necessary expertise to do the job properly. Before you hire a designer, find out what experience he's had. The Underground Space Center at the University of Minnesota (address below) maintains a list of professionals with expertise in earth-sheltered design. The advice of such experts is probably your best guarantee of a trouble-free house—*Charles A. Lane.*

Illustrations by Gene Thompson

ADDITIONAL INFORMATION

Following are some of the companies that offer plans for earth-sheltered houses: **American Solartron Corp.,** Rte. 5, Box 170, Centralia, IL 62801; **Andy Davis Caves,** Box 102, Armington, IL 61721; **Cave Enterprises,** Box 8283, Ann Arbor, MI 48107; **Earth Castle Homes,** Box 41702, Indianapolis, IN 46241; **Earth Shelter Corp. of America,** Rte. 2, Box 97B, Berlin, WI 54923; **Earth Systems,** Box 3533, Phoenix, AZ 85069; **Everstrong Inc.,** R.R. 3, Box 55B, Highway 19 and 71 East, Redwood Falls, MN 56283; **Simmons and Sun,** Box 1497, High Ridge Shopping Center, High Ridge, MO 63049; **Solar Earth Energy,** 2020 Brice Road, Reynoldsburg, OH 43068; **Solar Efficient Earth Dwellings Design,** 1027 E. Winona Avenue, Warsaw, IN 46580; **Terra-Dome Corp.,** Oak Hill Cluster, #14, Independence, MO 64057; **U'Bahn Earth Homes,** 4008 Braden, Granite City, IL 62040.

The following resource centers offer information on earth-sheltered construction: **Charles A. Lane Associates,** 3431 Kent St., Suite 909, St. Paul, MN 55112, (612) 482-9115; **The Underground Space Center,** 221 Church Street S.E., Rm. #11, Mines and Metallurgy, University of Minnesota, Minneapolis, MN 55455, (612) 376-1200; **The Clearing House for Earth-Covered Buildings,** School of Architecture and Environmental Design, University of Texas at Arlington, Arlington, TX 76019, (817) 273-3083; **Walter T. Grondzik and Lester Boyer,** School of Architecture, 103 Architecture Building, Oklahoma State University, Stillwater, OK 74074, (405) 624-6043; **Nolan Aughenbaugh,** Department of Mining, Petroleum, and Geological Engineering, University of Missouri, Rolla, MO 65401.

The following books, periodicals, and reports explain earth-sheltered design: *Earth-Sheltered Housing Design, Guidelines, Examples, and References* (available from the Underground Space Center, address above); *The Builder's Manual for Earth-Sheltered Construction,* by Charles A. Lane and Brent Anderson, Concrete Construction Publications, World of Concrete Center, Addison, IL 60101; *The Underground House Book,* by Stu Campbell, Garden Way Publishing Co., Charlotte, VT 05445; "Proceedings of the Earth-Sheltered Building Design Innovations Conference," School of Architecture, Oklahoma State University (address above), April 1980; "Proceedings: Alternatives in Energy Conservation: The Use of Earth-Covered Buildings," University of Texas at Arlington (address above); *Earth-Covered Buildings and Settlements,* edited by Frank Moreland, Vol. II CONF-7805138-P2, July 1979, National Technical Information Service, U.S. Dept. of Commerce, 5285 Port Royal Road, Springfield, VA 22161; *Earth-Shelter Living* (bimonthly magazine), WEBCO Publishing, 479 Fort Road, St. Paul, MN 55102.

passive-solar under-floor heat bank

Mark Jones has done it, and Scott Morris can prove it.

In the harsh Rocky Mountain climate of northern New Mexico, architect Jones has achieved passive-solar perfection. His series of natural-convection houses using under-floor rock beds are heated entirely by the sun; they use no fans, pumps, or blowers; and yet they maintain almost constant temperatures day and night.

Solar engineer Morris monitored the performance of the prototype house during the colder-than-normal January and February of 1981. His findings: The passive-solar heating system delivered essentially 100 percent of the heat needed for the 2000-square-foot house, and did so with a temperature swing of only 4.2 degrees F. Many thermostats don't do that well.

Now that the first of these under-floor-rock-bed houses has proved itself, refined—and less costly—versions have sprouted in the hills around Santa Fe. And Jones has plans for an entire subdivision of passive-solar homes. "We're bringing solar into the mainstream for Mr. and Mrs. Average Homebuyer," he told me.

That has always been the Holy Grail of solar architecture. Previously, owners of passive-solar houses needed a measure of missionary zeal: They had to be willing to adjust their life styles to less heat, give up certain amenities (such as north-facing windows), and put some daily effort into operating the system. Not in Jones's houses. In fact, the family that bought the first under-floor-rock-bed house had no real interest in solar; they simply liked the design of the house.

The outstanding performance and architectural flexibility of Jones's houses both stem from his success in perfecting, through four generations of the design, a thermosiphoning solar system. Thermosiphoning is based on a simple fact: Hot air rises. If its path is confined in a well-sealed loop and some means is provided to take heat out of the air at one point in the cycle and put it back in at another point, a constant flow of air will develop, rising and falling without benefit of any other force.

In a thermosiphoning solar system, a solar collector puts heat into the air and a thermal sink of water or rocks takes it out. This sort of loop is convenient for heating houses because the energy in the thermal mass can be released when and where it may be needed.

Thermosiphon systems are not new with Jones, of course. The concept can be traced back to the designs of Edward Sylvester Morse in the 1880s. The modern evolution of such systems began in 1967 when a solar innovator named Steve Baer designed a solar chimney in Bernalillo, New Mexico. He followed that with other designs and, in 1972, built the ground-breaking Paul Davis house in Corrales, New Mexico.

The progression begins

"When I first saw the Davis house in 1976," Jones recalled, "I was struck by the logic and simplicity of the thermosiphon system. Since neither heat collection nor storage occurs in the living space, I realized it would be possible to reduce the large temperature swings that typify other kinds of passive-solar design."

Jones's first attempt to design such a system was derivative of Baer's work: a hybrid system in which a convective loop puts heat into a rock bin, then distributes it to the house through a fan-forced duct system.

"That first collector and rock bin were designed on the basis of horseback estimates and trial-and-error tests," Jones says. But as a first try the system was a success (Jones and his family still live in the house), and he was encouraged to attempt a house that would be entirely passive.

To do this, Jones adapted a proposal by Morris to use the rock storage for interior walls of a house. The thermosiphon system would run the same way, driven solely by the temperature difference in the loop, but the air would be routed through large under-floor ducts to several tall, thin cages filled with 3- to 7-inch-diameter rocks. Then, when room air cooled below the temperature of the rocks, heat would radiate through the plaster on both sides of the vertical beds.

The house was built using 93,000 pounds of rocks in the steel-framed beds. The convective loop worked well, and the radiant heat offered greater comfort than the forced-air heat of his first design. But there were problems, too. Because the duct lengths varied, some beds got hotter than others, so the house heated unevenly. Also, the cost of hand-filling the beds limited the design's commercial prospects.

New ball game

But now the stage was set for a real winner. Why not lay down the vertical beds, providing the whole house with a continuous radiant floor panel? At first the idea seemed to be a mere extension of the previous design, with just a few parameters needing adjustment. Instead, it turned out to be a new thermal-mass concept.

The idea required that the rock bed extend under the entire 2000 square feet of floor area, which, because of the constraints of the site, would have to be designed along the north-south axis. Working out a rock-bed geometry to produce balanced air flows under these conditions proved extremely complex. Jones called in Morris.

The engineer's solution was the ingenious tree-shape duct system illustrated earlier. Unfortunately, building the 18-inch-deep bed and loading it with 110 tons of rocks took a huge amount of hand labor—three weeks' work for four men.

After the house was completed, Morris studied its performance under a contract from the Department of

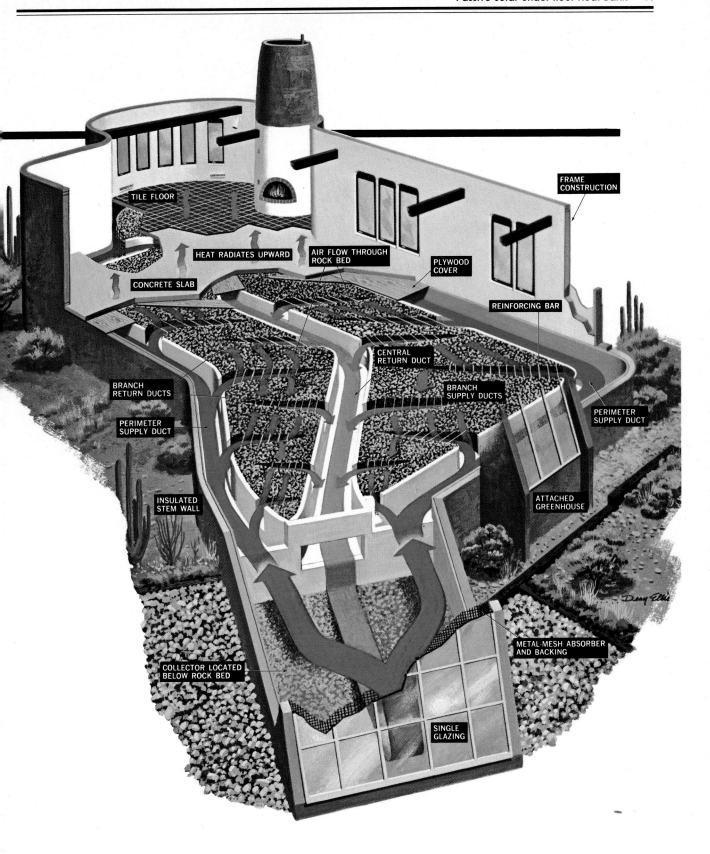

TILE FLOOR

FRAME CONSTRUCTION

HEAT RADIATES UPWARD

AIR FLOW THROUGH ROCK BED

PLYWOOD COVER

CONCRETE SLAB

REINFORCING BAR

CENTRAL RETURN DUCT

BRANCH RETURN DUCTS

BRANCH SUPPLY DUCTS

PERIMETER SUPPLY DUCT

PERIMETER SUPPLY DUCT

INSULATED STEM WALL

ATTACHED GREENHOUSE

COLLECTOR LOCATED BELOW ROCK BED

METAL-MESH ABSORBER AND BACKING

SINGLE GLAZING

First design of horizontal rock bed used 220,000 pounds of river rock in a tree-shape layout. Cool air rising through the 24 feet of collector height is heated by up to 70 degrees F, then enters duct system for heat transfer to rocks and return to bottom of collector. Large ducts and gentle curves compensate for low pressure of air flow. Average flow path through rock-bed sections is 4 feet (ample for good heat transfer), but path is shorter in parts of bed farthest from collector to allow for heat and pressure loss during passage through ducts. Thus, all parts of bed get equal thermal charge. Measured temperatures inside house showed variation from north to south rooms of less than 4 degrees, day-night swing in a given spot of about 3 degrees.

Energy. Its solar-heated fraction—excluding a small contribution from the attached greenhouse—worked out to be 97 percent even though the design goal had been only 85 percent. Similarly, the day-night temperature swings and north-south temperature variations were less than had been predicted.

Part of the reason for those astonishing numbers is that the walls of the house are better insulated and sealed than is usually the case with passive-solar houses that use adobe or concrete-block walls. In the under-floor-rock-bed design there's no need for large amounts of mass above the floor, only for a good thermal cocoon. The result is a house that looks like adobe, but is actually framed. It costs less to build and allows use of more insulation for a reduced heating load.

Some blocks replace some rocks

But if time and money were saved on the above-ground part of the house the hand labor of constructing the rock bed still dictated a selling price way up in the stratosphere. Jones recognized that his design would not be truly successful until it was adaptable to the production-house market.

To bring the cost down, he next attempted substituting concrete block for some of the rocks. The first of these designs used a tree layout similar to its predecessor, but the bottom 10 inches of rock were replaced by blocks laid loosely on their sides so their hollow cores provided air-flow passages through the bed. Above the blocks a layer of rocks was used, but these, it turned out, could be placed by an or-

Huge number of man-hours went into laying thermal mass of rock (top). Follow-up design cut costs. House is designed in Southwest style, but walls are framed—not adobe. Fireplace is an amenity only. Siphon collector must be below rock bed (bottom), so south-hill site is needed. Stack vents collect heat in summer.

This second-generation house (shown above left) employs rock-bed heat bank and is sited to overlook downtown Santa Fe. Architect Mark Jones' original personal home (right and below) demonstrated that a rock bed could be charged by natural convection. The bed is not distributed under the floor. Its warm air is distributed by a fan. Below are collector and pool details.

dinary backhoe at a great saving of time and labor."The block gives us a large amount of mass for a small amount of dollars," Jones said.

Two of those second-generation under-floor-rock-bed houses were built.

Blocks replace all rocks

The third, and newest, generation has some radical design changes. The thermal storage bed now has no rocks. The entire bed is made of concrete blocks lying on their side so that the cores form the air-flow passages. A heavier floor slab is placed directly on the upper course of blocks. And a sheet of plastic film on top of the blocks keeps things tidy during the pour. Extra reinforcing steel keeps the slab

crack-free during thermal cycling.

The mass/collector ratio has been adjusted to provide a better design balance, Jones says: "I think we had more than a little overkill in the mass ratio on the original layout, a lot more rocks than we really needed." The basic collector design is unchanged, but there are a lot of new details.

The glazing system has changed from redwood to an all aluminum structure with Neoprene seals. The night dampers that shut off any backdraft losses are now custom-built in Jones' shops with aluminum-faced insulated sandwich panels and foam gaskets. The whole assembly is operated by a sophisticated solid-state control module.

Yet the basics of the original design still apply: all natural convection, no fans. The air flow through the collector/storage system is completely isolated from the house, and the living space is a well-insulated cocoon. All thermal mass is contained in the block bed and floor slab.

At this point Jones "feels pretty comfortable with the design. It is well balanced, and most of the earlier construction complexities have been resolved. It's a lot simpler to build, more economical; we like it." Three of these houses have been built, and another is scheduled for construction as this yearbook goes to press—*Benjamin T. Rogers.*
Illustration by Dan Ellis.

log homes aren't just cabins anymore

Impossible was the word my friend used. "I don't know a soul who is paying under $300 a month for the mortgage on a new house," he said. "Our fathers may have paid less than $300, but times have changed."

I admitted that house prices, mortgages, interest rates—everything—had been troublesome in recent years. But I explained that $218.08 a month could conceivably be what he'd pay to live comfortably in a three-bedroom, one-bath, 960-square-foot log house. That figure represents the monthly charge for principal and interest on a $20,000, twenty-five-year mortgage at 12½ percent (today's going rate).

Is there a catch? Not if the buyer is willing to put in his own labor, and assuming he already owns a small parcel of land. Here's how:
● Precut package (logs are precision-cut, predrilled, and numbered for walls. Package also includes posts for foundation, all other structural parts, spikes, foam sealant, caulking, trim, windows, and doors). Cost: $20,000.
● All other materials and equipment (purchased locally) to complete the house (at 1983 prices). Cost: $7,000.
● Foundation (logs for post foundation included in precut materials). Cost: $0.

● Labor (by owners, family, friends). Cost: $0. (This isn't quite true, because buyers do spend some money, even if family members pitch in; there are extra meals, for example, and perhaps a case of beer now and then. Trade-offs are another way to get "free" labor; an owner might, for example, trade the muscle-power to cut and deliver several cords of firewood for the services of a friend who just happens to be a licensed plumber.)
● Land (previously owned.) Cost: $0.
● Utilities (wiring and plumbing—enough to run house lines into the ground—are included in materials purchased locally). Cost: $0.

The only costs, then, are $19,500 for the precut package and $7,000 for the locally purchased materials. That brings the total to $26,500. With a $6,500 down payment, a $20,000 mortgage is required, assumed, here, to be a twenty-five-year loan at 12½ percent.

The log house described is the AMP Econ-O-Plus, developed by Appalachian Log Structures. "AMP" stands for "Affordable Mortgage Payment." The company also offers a two-bedroom, one-bath, 720-square-foot model for a package price of $14,100 and es-

timates materials cost at about $6,000 for a total of $20,100.

While these prices may seem unusually modest, they are consistent with packages offered by most log-house manufacturers. The companies also produce many precut models containing between 1000 and 2500 square feet for package prices between $15,000 and $30,000 and finished costs in the 50s and 60s. Some log-house firms have been involved with homes costing over $500,000.

Once up and properly sealed, log houses deliver another saving—energy. "We were very pleasantly surprised," said Hal Dunham, an executive with Edison Electric Institute in Washington, D.C., referring to his utility bill. "Our two-year-old log home is all-electric, and bills for December of last year and January of this were only $252 per month. Owners of neighboring frame houses with less than our 3500 square feet were paying that much just for heating oil."

Wood is a natural insulator and thick wood is a heat bank, absorbing heat (solar or man-made) during the day and releasing it during cold winter nights. Most log walls are from 6 to 8 inches thick, and some are as much as 12 inches thick. None of these walls

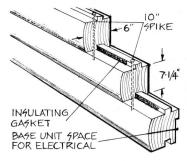

Most log houses are sold for use as year-round homes. Home below is from Ward; the one above is from Southern Structures. Indoors view, above right, model from Lincoln Logs, Ltd., Chesterton, NY, features flat-cut logs for easy furniture placement. Drawing and photo at right show log-joining techniques.

require additional insulation. In fact, log walls represent another saving—time. A frame wall begins on the outside with siding; then there's sheathing, framing, insulation, vapor barrier, plasterboard, taping, spackling, and either paint or wallpaper. A log wall, stacked once, is the finished exterior and interior, insulation and structural support, all in one.

Roy and Karen Ruddle of Hurricane, West Virginia, embarked on construction of their own log home last September, and it was ready for occupancy in mid-April. "Our log walls went up easily," says Karen. "Knowing they would helped us make the decision to buy a log-house package. Finishing the inside, though, went more slowly than we expected, but then we had never done this before."

Was building their own log house worth the effort? "At first, we didn't think we could afford a loan for a house because the interest rates are so high," says Karen. "But then we discovered that log-house prices are more reasonable, and we both like the rough style (or maybe I talked Roy into it). Now that the house is com-

pleted, we know it was worth all our effort (and the labor of our parents, brothers and friends). We invested about $40,000 in house and land; we know we saved at least $20,000 and Roy is convinced we could get $75,000 if we sold tomorrow."

Would they do it again? "Oh, yes," says Karen. "Staining shelves, pounding nails, and working on the yard are wonderful ways to take away the pressure of our careers. I teach mentally retarded high-schoolers, and Roy is a researcher in process safety at Union Carbide."

Like most companies, Lincoln Log Homes of Kannapolis, North Carolina, offers precut logs in a package, ready for do-it-yourself assembly. But Lincoln also offers a faster way to erect walls. Its logs are prebuilt into wall sections at the factory, fastened with lag bolts under high compression. Then sections are shipped to the site and put in place over a foundation with a crane. The shell of one such house was erected by a company crew in six hours, forty minutes.

Most manufacturers precision-cut their logs on huge power-driven saws.

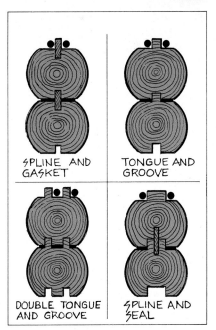

End views of log walls illustrate four typical methods for stacking log on log. Note that all the methods use some form of tongue or spline to assure an airtight fit. The joint usually has a foam gasket and is well caulked.

Spirits rise after the foundation is completed and the logs delivered. The forklift (left) is removing logs for an Appalachian Structures home for Roy and Karen Ruddle in West Virginia. The walls go up in a very short time. The crew (middle left) includes Roy and some relatives and friends. Photo (middle right) shows crane lifting logs from Lincoln Logs, Ltd., into place. Bottom photo is of Lincoln Logs ranch.

Most offer debarked logs left round outside, and many give you a choice of round or flat log walls inside. Others remove bark and shape the felled tree with a draw knife, straight hoe, and other hand tools. For example, Southern Rustics sells only hand-hewn whole logs, nearly always erected over a foundation by a professional crew using a boom or crane truck. Openings for doors and windows are cut later.

Eugene Davis, vice-president of Southern Rustics, points out that log houses lend themselves to do-it-yourself completion because there is much less to be done on the outer walls than must be done on a frame home. "But I see the real savings for a log house in reduced energy use year after year, low maintenance costs, and appreciation of one's investment. A log house does not deteriorate and lose value," he said.

Rick Steelman, president of Cedardale Homes, is even more enthusiastic. Says he, "Not only can you build a log house for 30 percent less than it would cost you to build a comparable home by conventional methods, but once you move in, it will cost about 30 percent less to heat and cool your house."

Doris Muir, editor of *Log Home Guide* and the moving force behind the newly formed North American Log Builders Association, lives in a log house she built on her own and is convinced that "log houses will last an average of 500 years . . . if built properly and foundation and roof are cared for (properly)."

Estimates of the number of buyers who complete their own log houses or act as their own general contractor (a potential saving of 20 percent over a contractor price) ranges from 50 to 75 percent of all buyers. As this is written, industry estimates indicate 1982 sales at around 15,000 log houses. All companies contacted reported a rise in the number of do-it-yourselfers buying log house packages. "The buyers who completed our log houses on their own have represented 50 to 60 percent of our business for a number of years," said Wilbert Bossie, director of mar-

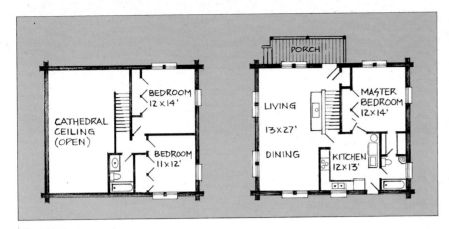

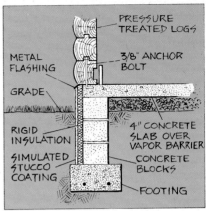

Styles are as numerous and flexible as those of conventional houses. Split level, from Appalachian Structures, has soaring cathedral ceiling on first level, as photo of customized model (left) shows. Plan demonstrates spaciousness within. Cross section (above) shows one method for tying logs to the foundation. Note the channel for wiring.

keting for Ward Cabin Co., oldest log-house package company on the continent.

Most buyers opt to finish their log house interiors in the same rugged style, using hardwood flooring, cedar plank paneling for partitions, natural wood steps and railings, and pine or cedar kitchen cabinets. Others complete their houses as one would a conventionally framed home—all a matter of taste.

Just as we are relearning and applying passive solar techniques discovered centuries ago, perhaps we would do well to take a much closer look at log-house advantages our forefathers knew. After living in conventionally framed houses all their lives the Dunhams were asked how they liked their two-year-old log house. Hal Dunham put it simply: "We love it!"—*John H. Ingersoll.*

ADDITIONAL INFORMATION

Write to the manufacturers. Addresses of firms mentioned in this article are: Appalachian Log Structures, Burke-Parsons-Bowlby Corp., P.O. Box 86, Goshen, VA 24439; Cedardale Homes, Inc., 400 Friendship Center, Greensboro, NC 27409; Lincoln Log Homes, 1908 N. Main Street, Kannapolis, NC 28081; Lincoln Logs, Ltd., Riverside Drive, Chestertown, NY 12817; Southern Rustics, Inc., P.O. Box 296, Foley, AL 36536; Ward Cabin Co., P.O. Box 72, Houlton, ME 04730.

Or send for a copy of the 1983 Winter Directory issue of *Log Home Guide* ($10 postpaid), which includes addresses and information on 137 U.S. and Canadian log-house manufacturers. Write *Log Home Guide,* Muir Publishing Co., Ltd., 1 Pacific Avenue, Gardenvale, Que., Canada H9X 1B0.

Also, write for info on 22 log cabin kit makers, that are members of the Log Homes Council, part of the National Association of Home Builders, 15th & M Sts., Washington, DC 20005.

concrete prefab homes

Would you buy a $33,000 Beaujolais from a door-to-door salesman? Or a Chablis or a Cognac? If you lived near here, Clearwater, Florida, you could. And if you were in the market for a house that cost up to 20 percent less to build than a conventional house, and used 30 percent less energy, you'd probably be wise to make the purchase.

The 1400-square-foot, two-bedroom Beaujolais is one of twenty-five models a new-to-the-U.S. French prefab-home builder now offers in this Sunbelt city. Maison Phenix homes use a unique concrete-and-steel system developed and used in Europe since shortly after World War II. The system permits rapid, on-site assembly by a small crew, Gerard Boudin told me. Boudin, a young, energetic Frenchman, manages Maison Phenix's U.S. operation.

"Everything starts at the unit construction plant," says Boudin. That's where the concrete panels are manufactured. Materials are trucked to the building site, where they're assembled by a Maison Phenix-trained crew (see photos). "We guarantee completion within three months," says Boudin, "but we actually complete most homes in two months."

Where conventional construction might use from twenty to thirty subcontractors to build a house, Maison Phenix uses only ten. That and the rapid assembly produce most of the savings.

Because of the concrete-panel construction, the homes look a little like bunkers before the finish stucco is applied. "But people don't complain," says Boudin. "It doesn't look nice, but it does look *strong*!" In fact, the homes exceed U.S. requirements for hurricane resistance. Of course, once stuccoed, Maison Phenix homes look no different from any other.

Under the stucco, however, are important differences. Exterior walls have three major elements (see inset): concrete panels, steel framework, and drywall attached to 2 × 2s. "It works like a thermos bottle," explains Boudin, "except, of course, without the vacuum." The three elements have very few direct connections, and in between there's a 2-inch air space. It all produces an R-19 wall with very low air infiltration.

Maison Phenix now has two divisions operating in the Tampa-St. Petersburg area. (Each division, with its own unit construction plant, covers an area based on a maximum 1½-hours driving time to home sites.) The homes are sold on a door-to-door basis. "We don't wait for customers to visit our models," says Boudin. By 1986 he plans to cover all of Florida selling 5000 homes annually.

Boudin isn't missing any bets on reaching his 1986 goal: Maison Phenix offers models ranging from the 900-square-foot Anjou, at $27,000, up to the 7000-square-foot Atrium, priced at $400,000. Something for everyone, even in Palm Beach—*Richard Stepler.*

Anchor bolts embedded in concrete slab secure steel framework for outer walls. Each wall includes a St. Andrew's cross for strength (two are visible in photo).

Framework complete, precast reinforced-concrete panels are bolted in place. Panels measure 32 by 48 inches to allow two workers to handle them.

Wooden roof trusses are double-bolted to steel framing with tie-down bars for hurricane resistance. Three-person crew can erect shell in one week.

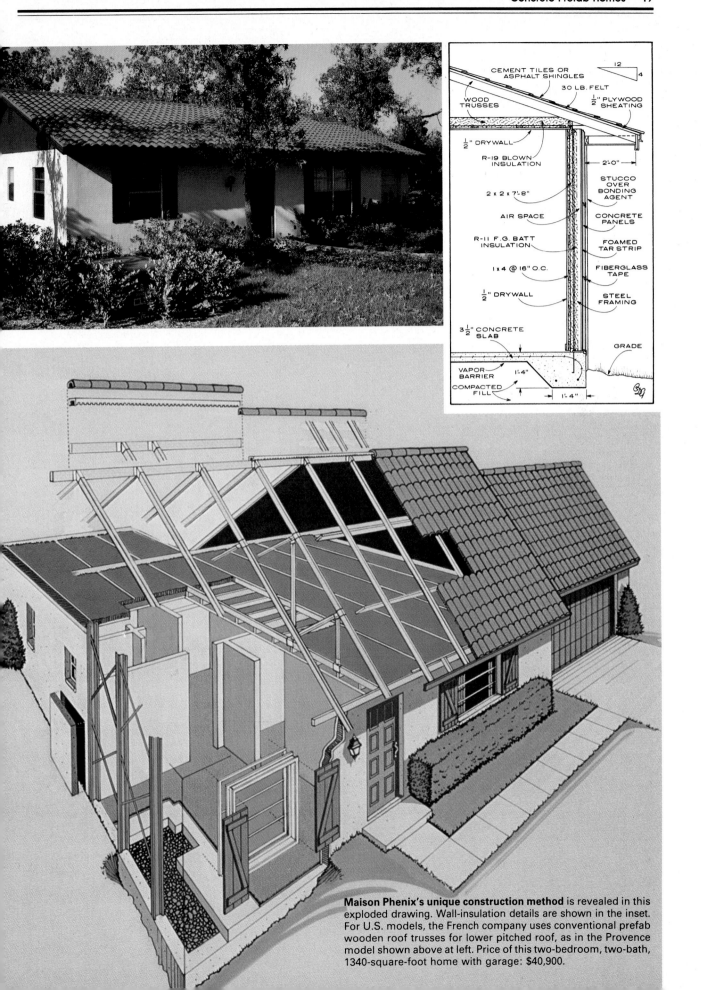

CEMENT TILES OR
ASPHALT SHINGLES

30 LB. FELT

WOOD
TRUSSES

½" PLYWOOD
SHEATING

½" DRYWALL

R-19 BLOWN
INSULATION

2'-0"

STUCCO
OVER
BONDING
AGENT

2 x 2 x 7'-8"

CONCRETE
PANELS

AIR SPACE

FOAMED
TAR STRIP

R-11 F.G. BATT
INSULATION

FIBERGLASS
TAPE

1x4 @ 16" O.C.

STEEL
FRAMING

½" DRYWALL

3½" CONCRETE
SLAB

GRADE

VAPOR
BARRIER

1'-4"

COMPACTED
FILL

1'-4"

Maison Phenix's unique construction method is revealed in this exploded drawing. Wall-insulation details are shown in the inset. For U.S. models, the French company uses conventional prefab wooden roof trusses for lower pitched roof, as in the Provence model shown above at left. Price of this two-bedroom, two-bath, 1340-square-foot home with garage: $40,900.

adding a solar room

Northeast winters tend to be cold and gray. But the sun-drenched room I designed and added to my home has changed that. It's a warm, bright add-on, suitable for use as a breakfast room, for simply relaxing with a book, or for hot-tubbing in a whirlpool tub I've installed. As I write this in midwinter, a small fan is quietly blowing 80-degree-F air from the sunspace into my house.

This project called for inexpensive, readily obtained materials and do-it-yourself building techniques, since I planned to do most of the construction. Also, I wanted to use the space both summer and winter, but with minimal shading or insulation.

A 60-degree sloped bank of windows is ideal for capturing the winter sun. However, combine that with a vertical knee wall (for ventilation) and a sloped roof (to shed water), and you have an unstable configuration. To solve this problem I designed a structural system of prefabricated plywood ribs. Located at the glazing joints, they provide both roof structure and window mullion.

An insulated roof helps keep the sunspace warm in winter and cool in

Snug interior of add-on sunspace during winter months is aided by fiberglass insulation for the floor and ceiling, plus moderate thermal storage mass to capture solar energy streaming through large, sloped windows during the day. The outdoor platform for a 195-gallon whirlpool spa (middle, far left) sits on a leveled gravel bed. During winter months, this portable Jacuzzi can be moved inside the sunspace (bottom, far left). Architect Milstein decided to glaze one end wall for a nice view (near left), leaving other wall solid except for a low awning to catch westerly summer breezes. Quarry-tile floor over concrete stores solar heat.

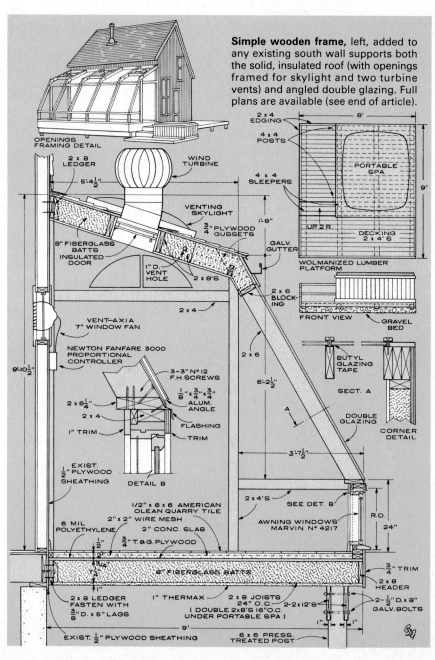

Simple wooden frame, left, added to any existing south wall supports both the solid, insulated roof (with openings framed for skylight and two turbine vents) and angled double glazing. Full plans are available (see end of article).

Pouring and smoothing cubic yard of concrete took an hour. After striking off excess flush with form, finish with bull float.

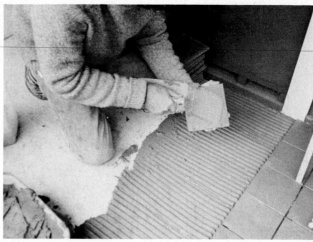

Half-inch quarry tiles atop 2-inch concrete slab create thermal storage mass. Bed them in a thin-set mortar.

Ribs are formed with plywood gussets glued and nailed across the joint on each face. Use waterproof glue such as resorcinol.

Inside finishes for the sunspace are ½-inch gypsum board on the wall and ceiling. A 6-foot-wide door opens into the house.

summer. A 2-inch-thick concrete floor slab covered with ½-inch quarry tile provides some thermal mass to store solar heat. To maximize solar heat for direct warming of adjacent areas of your house, you should keep thermal mass to this minimum. But to retain most heat *within* the sunspace—say, for growing plants—you may want a greater storage mass.

Natural ventilation for summer is important. Awning windows are installed along the floor perimeter, and two 14-inch-diameter wind turbines and a venting skylight are on the roof. Insulated, weatherstripped doors were built to seal off the turbine vents in the winter. On winter nights, I close the double French doors between the house and sunspace, allowing the outer temperature to drop into the forties.

To control the sunspace environment, I've installed a Vent-Axia window fan in the wall between the sunspace and the house. A special controller for the fan has a low setting

that protects plants from freezing by blowing heat from the house on unusually cold nights. Another setting blows warm air into the house on sunny days. A solid-state circuit automatically varies fan speed with demand.

A 10-foot × 10-foot × 8-inch concrete-block Trombe wall existed on the side of my house before I began the sunspace. After removing the glazing, this also became thermal storage mass. Unless you have a masonry house, a similar heat storage probably isn't available to you. But the equivalent thermal storage is possible with about twenty-five 20,000-Btu phase-change tubes placed in direct sun against the wall. Thirty cubic feet of water in containers against the wall can achieve the same mass.

Selecting the glazing material was easy: The 46 × 76-inch (or 34 × 76-inch) insulated, tempered, sliding patio-door replacements I used are the most economical double glazing I could find. I

bought them for less than $80 each by shopping around. Although low-iron glass would transmit more light into the sunspace, I found it was expensive and hard to find—*Jeff Milstein. Drawing by Carl De Groote.*

PROJECT SPONSORS—AND WHAT THEY SUPPLIED

American Olean Tile Co., Lansdale, PA 19446 (6 × 6-inch Canyon Red quarry tile); **Celotex Corp.**, 1500 N. Dale Mabry Highway, Tampa, FL 33607 (Thermax insulation); **Crown Vent**, Dundee Park, Andover, MA 01810 (Vent-Axia fan); **Jacuzzi Whirlpool Bath**, P.O. Drawer J, Walnut Creek, CA 94596 (Cambio portable spa with redwood skirt); **Koppers Co. Inc.**, Pittsburgh, PA 15219 (Wolmanized lumber for Jacuzzi platform); **Marvin Windows**, Warroad, MN 56763 (awning windows); **Newton Electric**, 2390 River Road, Selkirk, NY 12158 (proportional controller for fan); **Olympic Stain**, Box 1497, Bellevue, WA 98009 (exterior stain); **Sears, Roebuck & Co.**, Sears Tower, Chicago, IL 60684 (14-inch-diameter wind-turbine vents; skylight; 5-foot sliding patio door with storm glazing).

HOW TO ORDER YOUR PLANS

The accompanying sketches are incomplete and are only offered to help you evaluate the project. To order large-format DIY plans, including clear construction details, sources for materials, plans for an insulating night shade, and building hints, send $15 to: Jeff Milstein, Box 413, Dept. PS, Bearsville, NY 12409.

converting attics to living space

Many attics offer enough head room to allow room additions without the added cost of raising the roof or extending outward from the existing foundation. Basically, roof construction is either by trusses or by conventional framing. A trussed roof consists of framing members, usually wood, constructed in a manner that leaves little extra space. Expansion into an attic with a trussed roof is almost always too expensive. Framing members can't be removed without weakening the roof structure, and nothing should be done without first consulting an architect or structural engineer.

A conventionally framed roof, on the other hand, can often be used to gain additional space. Normally this type of roof consists of no more than ceiling joists over the living space, rafters for the roof, and collar beams overhead every third or fourth rafter. Occasionally there are additional framing members—wood braces—used to cut down the span of rafters and, therefore, reduce their size. If your attic has these additional braces, design your expansion around them—they should not be removed without expert advice.

Is there enough space?

The generally accepted minimum headroom for any living space is 7 feet from floor to ceiling. A lower ceiling makes an area seem cramped and uncomfortable even to people under 6 feet tall. In fact, if your family members are exceptionally tall, you may even want to go to an 8-foot ceiling height.

Normally, an attic has 7 or 8 feet of headroom—and some have even more—near the roof ridge. The problem arises in the width of the area in which you can maintain the desired headroom. A 12-foot area can accommodate almost any use you might want to put it to, and the wider the area, the more versatile the design possibilities. If the usable width is less

than 12 feet, you're severely restricted in how you can use the space and, if it is only a few feet wide, the attic is useful for little more than storage, unless you decide on the expensive approach of raising one or both sides of the roof.

To good advantage

There are four basic ways to take advantage of an attic space without disturbing the roof.

1. Build the side walls to standard room height and add a ceiling overhead. A ceiling will make the room feel like a standard room, since the sloping attic lines are lost, but this works only if the attic area is spacious.

2. Build the side walls to standard height and leave the top open to the rafters. This approach is suitable for attics that are not so wide—the openness overhead makes the space seem

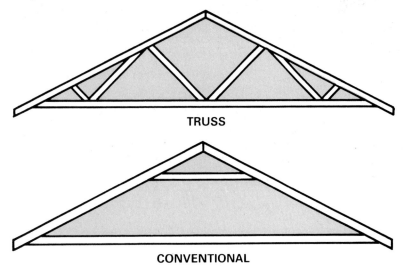

TRUSS

CONVENTIONAL

Trussed roof obstructs the use of an attic for living space. In most cases, a conventionally framed roof allows easy conversion of attic to living space.

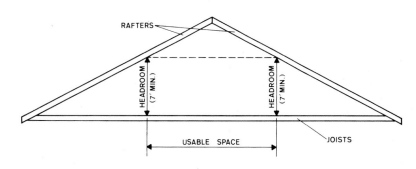

To determine how much of attic area is usable for living space, measure the width of the area in which you are able to maintain a minimum headroom of 7 feet.

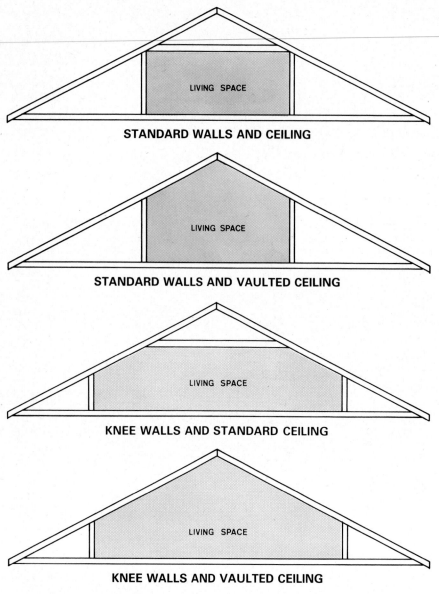

STANDARD WALLS AND CEILING

STANDARD WALLS AND VAULTED CEILING

KNEE WALLS AND STANDARD CEILING

KNEE WALLS AND VAULTED CEILING

The four ways to develop the shape of an attic: Side walls may be full height or shorter knee walls. And in either case, a standard flat ceiling can be used, or area can be left open to rafters for a more spacious feeling.

larger. Also, vaulted ceilings may have special design appeal.

3. Build shorter "knee walls" at the sides, instead of standard height walls, and add a ceiling. This will add to the size of the room, and the low ceiling space along the knee wall can be used for furniture, seating, and other uses that don't require full headroom.

4. Build knee walls and leave the top open to the rafters. Whether or not you install a ceiling is more a matter of taste than cost factor, but again, the room that's open above will seem larger than it is.

The generally accepted height for knee walls is 4 feet. But this standard was established when roof slopes tended to be steeper than they are today. A better way to determine the height is to locate the point at which

Before starting work in attic, put in temporary footing (plywood or similar material), securing it to joists. Place edges near a joist so you don't overstep and tip over the footing material.

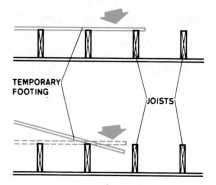

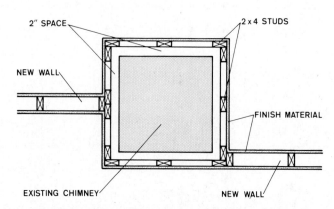

If chimney must be enclosed (above), maintain 2-inch spacing between it and framing materials. To reduce wasted space, try to combine it with a new wall system.

the headroom is 6 feet, or equal to the tallest member of the family over 6 feet, and then move 3 feet toward the outside wall. For a roof pitch of 8 in 12 (8 inches of rise per foot of width), the knee wall will be exactly 4 feet. A flatter roof will yield a taller knee wall. If, however, your roof is steeper, come back to the point where the knee wall will be 4 feet high. A shorter knee wall will make the area difficult to clean. Also, the 4-foot height makes installation of finish materials easier.

Which rooms?

An attic can be converted into almost any kind of facility, with a few notable exceptions. It's not a good idea, for example, to plan the area for use by elderly or handicapped persons who might find the stairs a problem. And don't plan the space as the main kitchen and family/great room. These are often-entered rooms and the location would create traffic problems and overburden the family with stair climbing. Also the outside access that is often an essential feature of a family/great room or kitchen is not available in an attic location.

Ideal uses for attic space include: standard bedrooms, master bedroom suite, guest bedroom, bath, recreation/ entertainment room, adult retreat, den, office/study, hobby/craft center,

Through-attic chimney is expensive to move, but it can be turned into the focal point of the new living area by using it as a room or space divider. A hearth can be added for cushions and seating, or to support a wood-burning stove or a free-standing fireplace.

library, stereo/music room, darkroom, artist's studio, playroom.

Plumbing in an attic

Since the attic is above the main level of the house, there is no problem getting the waste lines to drain by gravity flow, leaving you free to design any type of plumbing you wish. The only rule is that the attic walls containing the plumbing should be directly over the walls that contain plumbing in the floor below. This will greatly simplify and reduce the cost of plumbing hookups. If you want to install two baths, or a bath and kitchenette or wet bar, make the rooms back to back so you can use one connection to existing plumbing.

Adding natural light

Most attics are dark, having only a small gable window at each end—or no window at all. This is a relatively easy problem to solve; you can add skylights, gable windows, and/or dormers.

Skylights are the least expensive way of getting natural light into an attic room and can be installed with a minimum of labor. They can be installed at almost any location, making them the most versatile solution to the natural-light problem. Openable skylights have the further advantage of allowing room ventilation and a measure of temperature control. In warm weather, an openable skylight near the ridge will let hot air rise out

of the room, while cool air enters through a lower window.

Gable windows are more limited— they can only go in gable walls. If you have a hip roof, you can't install a gable window. If you already have a window in one or both of your gables, it's probably too small for your attic conversion. Remove it and install a new, larger window, or add extra windows on each side.

Dormers are a little more difficult to install; they require some roof removal, a lot more framing, and a new roof area. They do, however, give the attic area extra space, though usually not enough to make a real difference. They also affect the house's architectural appearance; for this reason, if dormers are part of your plan, you may want to add them to the rear of the house. If you add them on the front, put in at least two, equidistant from the house ends or otherwise balanced with the present design.

Chimneys, ducts, pipes

One or all of these are likely to be rising through your attic. Vent pipes and ducts can usually be enclosed as part of a new partition, or moved, if necessary, without too much difficulty. If the vent is a plumbing stack, it probably rises directly above the lower floor's plumbing wall, so it can be located within any new plumbing wall needed for the attic conversion.

A chimney is a tougher problem—

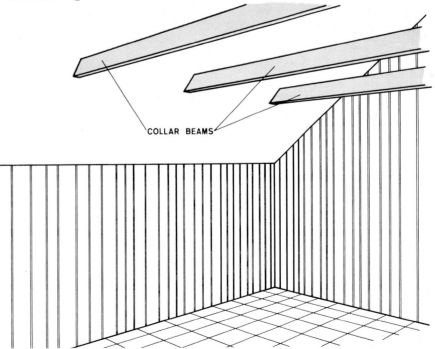

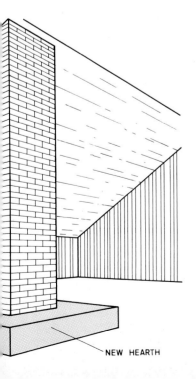

NEW HEARTH

Collar beams can be raised a few inches to gain room; but do not remove them without first consulting an expert. Instead, leave the collar beams exposed (above), and then, trim them out with finishing lumber for an interesting ceiling effect.

COLLAR BEAMS

it costs a lot to move one. So, if possible, turn it into an asset by using it as a room divider (paint the brick, if you wish) or putting a wood stove or freestanding fireplace in front of it. If the chimney brickwork is too ragged to make it the focal point of a room, you can frame around it and cover the frame with finish materials. Keep the wood framework at least 2 inches away from the chimney and, if possible, work the enclosed chimney into any wall system you plan.

Adding a stairway

Another problem that must be considered when you plan an attic conversion is where to locate the stairs. Access to many attics is simply through a square hole in a hallway or closet ceiling, or up a folding staircase. Of course, neither of these is sufficient; if access to attic living space isn't reasonably convenient, the area will not be used. Try to install a standard staircase or, if space is very tight, a spiral staircase; nothing less will suffice.

The location of the stairs is usually dictated by the design of the main floor. Wherever the staircase can be placed on that level sets where it will enter the attic. But you should also consider the attic development. If the staircase enters in the middle of the attic, it makes a natural division for two rooms. If it must be located at one end of the attic, you may be limited to a single room, unless you have enough width for a hallway. The important thing is to plan your remodeling so the stairway does not begin or end in a private room—such as a bedroom—that family members must pass through in order to get to shared space.

Many municipalities require two exits for attics used for living space (especially bedrooms). Even if you're not required to, you may want to add an additional escape route for your family in case of fire. If you don't have room for two sets of stairs, get an emergency ladder and keep it near a window large enough to crawl through. An emergency ladder doesn't cost a lot, and it will add tremendously to your peace of mind. Just be sure every family member fully understands how the emergency escape is used.

Construction problems

Most attics are poorly ventilated and can become extremely hot. Before the area is insulated and a cooling system installed, temperatures can reach more than 140 degrees. That's why it's a good idea to avoid working in the heat of summer; even the young are susceptible to heat stroke. Also, it's easier to hire any subcontractors you may need in cooler weather—sometimes even at a better price than during their busy summer months. And make ventilation your first construction priority; skylights, windows, and vents should be installed as soon as you've laid temporary footing. Wear a breathing mask when doing any work, such as sawing, that creates dust. Don't take safety lightly, especially when working in a poorly ventilated and poorly lighted attic.

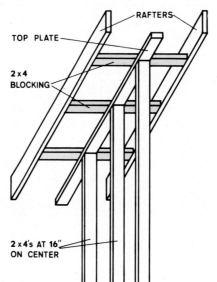

For walls that run parallel to the rafters (left), provide braces at the top, between the rafters, for added stability.

If flat ceiling is put in over attic rooms (below), it should be constructed in same way as a standard ceiling, with 40-pound live-load capability, full insulation, an access hole, and proper ventilation.

Getting large materials into the attic may also be a problem. If you can't hoist materials up from the outside through a window or other opening, you may have to cut them to size before transporting them to the attic.

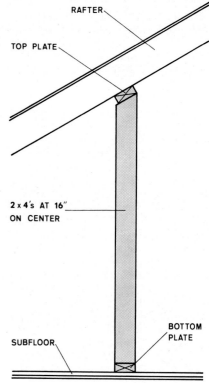

For walls perpendicular to rafters (above), studs should be cut on the same slope as the roof at the top, with the top plate secured to the rafters as shown in drawing.

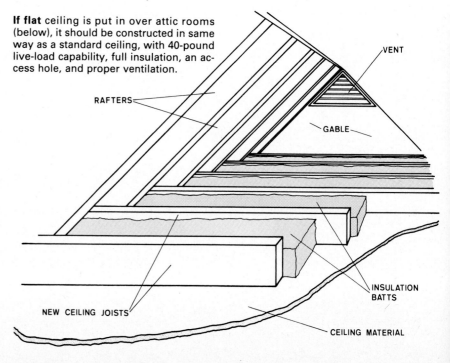

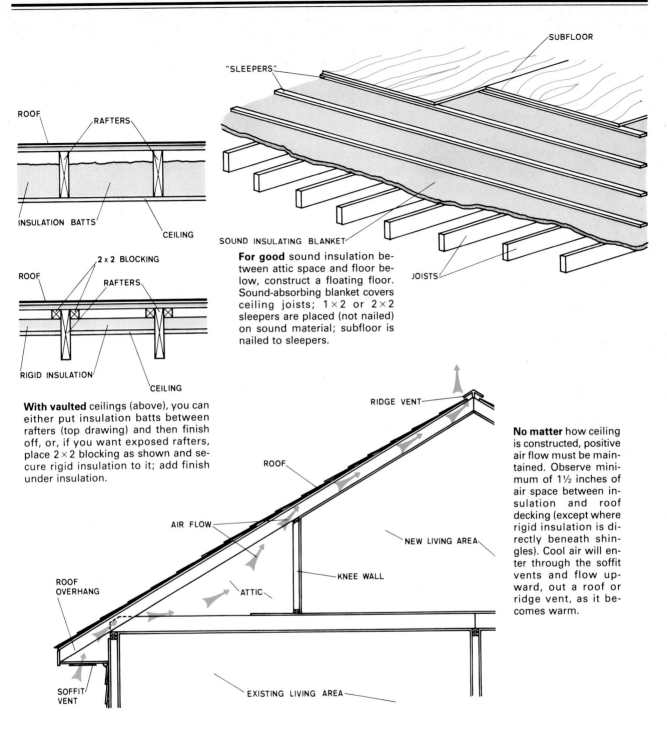

ROOF — RAFTERS

INSULATION BATTS — CEILING

2 x 2 BLOCKING

ROOF — RAFTERS

RIGID INSULATION — CEILING

"SLEEPERS" — SUBFLOR

SOUND INSULATING BLANKET — JOISTS

For good sound insulation between attic space and floor below, construct a floating floor. Sound-absorbing blanket covers ceiling joists; 1×2 or 2×2 sleepers are placed (not nailed) on sound material; subfloor is nailed to sleepers.

With vaulted ceilings (above), you can either put insulation batts between rafters (top drawing) and then finish off, or, if you want exposed rafters, place 2×2 blocking as shown and secure rigid insulation to it; add finish under insulation.

RIDGE VENT

ROOF

AIR FLOW

ROOF OVERHANG

SOFFIT VENT

NEW LIVING AREA

KNEE WALL

ATTIC

EXISTING LIVING AREA

No matter how ceiling is constructed, positive air flow must be maintained. Observe minimum of 1½ inches of air space between insulation and roof decking (except where rigid insulation is directly beneath shingles). Cool air will enter through the soffit vents and flow upward, out a roof or ridge vent, as it becomes warm.

But carefully record any measurements needed for cutting so you don't waste expensive materials.

Although most ceiling joists are designed to support a minimum "live" load of 40 pounds per square foot, which is sufficient for furniture and normal activities, there is a possibility that your ceiling joists may need to be beefed up. If you are unsure, have an expert look at your attic and calculate the bearing capacity of the joists. Should they need additional support, he can advise you of the best approach to take.

The small cost of an hour or two of an expert's time could save you from an expensive disaster later.

Sound insulation

A thick carpet and pad installed in the attic will absorb most low-frequency sounds, such as footsteps. For additional sound insulation, construct a relatively inexpensive "floating floor." Lay a full covering of insulation or other sound absorbing material over the existing joists or decking. Then place 2×2 sleepers at 16 inches on center, perpendicular to the joists.

Don't nail the sleepers—just lay them in place. Then nail the subfloor to the sleepers. This will greatly reduce all frequency levels of noise traveling from one floor to another.

If carefully planned, an attic conversion can add a lot of living space at minimum cost. Look at your attic closely before you begin any work. Determine the problems you will encounter, then develop a strategy for solving those problems. When you do, the work will not only go more smoothly, but, in the long run it will cost a lot less—*Herb Hughes.*

raising the roof to gain rooms

A typical A-shaped attic in a Cape Cod-style home built in 1942 offered an opportunity to create a dramatic new master bedroom and bath. However, to get the bright, airy bedroom and luxuriously sized bath shown on these pages, the owners had to work out design solutions to problems caused by the existing stairs, chimney, and plumbing. Their most significant decision, perhaps, was to raise the roof and build a dormer the full length of the house; in the end, this proved to be only slightly more expensive than the alternative of a much smaller dormer. The work progressed as follows:

1. The rear half of the roof was completely removed, rafters cut, cross-ties eliminated, and temporary bracing set under the front half of the roof.

2. Joists were extended for the dormer and balcony, new exterior walls framed, and a new roof and wall sheathing were applied.

3. New windows, sliding doors, and skylights were then installed.

4. The plumbing, wiring, and insulation were installed, the bath fixtures were set, and the balcony deck was completed.

5. The chimney was framed with a new wall and a closet was built on the north side of the chimney.

6. Next step was to drywall the interior.

7. New shingles were applied over the old roof and rolled roofing was installed over the dormer.

8. Finish work—painting, trim, gutters, tiling—was completed.

A special feature of the design is that all potentially lost space (the space between the studs, for example) was put to work. In the bathroom, it was used for a towel shelf; on the west wall, for plant shelving and a window seat; on the closet (chimney) wall, it became built-in bookshelves, and, in the area at the head of the stairs, two small storage closets were created.

The owners achieved the three goals they had set for themselves in this project: They created an inviting living/sleeping space and a luxurious bath. They created a room bathed in natural light and open to a beautiful view. And they maintained the integ-

Before-and-after photos show a pleasing transformation from Cape Cod peak to full-length dormer, which greatly increased living space at relatively reasonable cost. Note recommendations for roof circulation in drawing on next page.

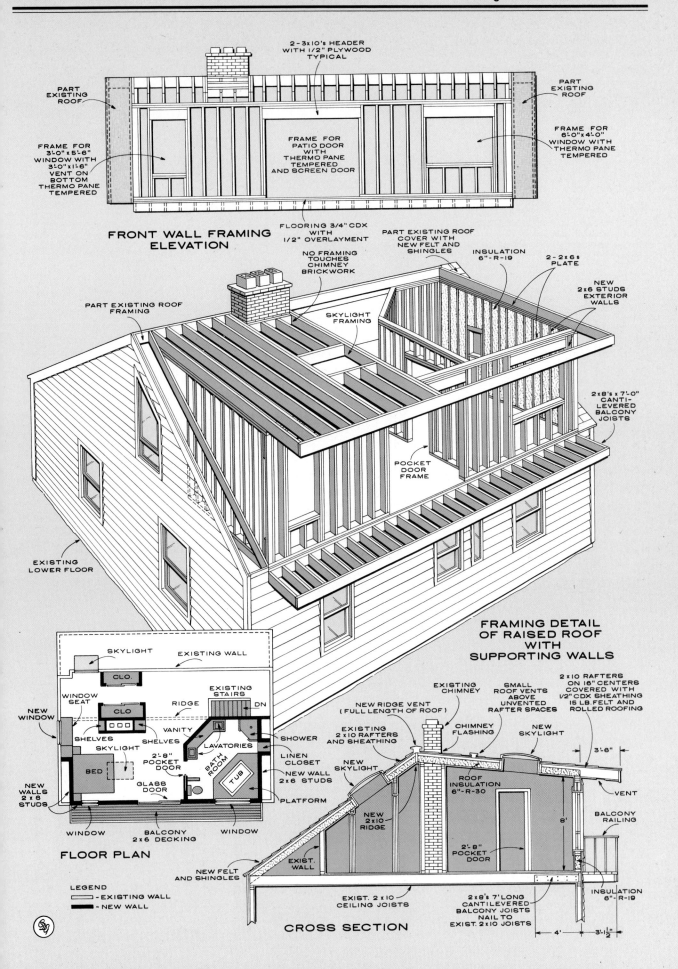

2-3x10's HEADER
WITH 1/2" PLYWOOD
TYPICAL

PART
EXISTING
ROOF

PART
EXISTING
ROOF

FRAME FOR
3'-0"x5'-6"
WINDOW WITH
3'-0"x1'-6"
VENT ON
BOTTOM
THERMO PANE
TEMPERED

FRAME FOR
PATIO DOOR
WITH
THERMO PANE
TEMPERED
AND SCREEN DOOR

FRAME FOR
6'-0"x4'-0"
WINDOW WITH
THERMO PANE
TEMPERED

FRONT WALL FRAMING
ELEVATION

FLOORING 3/4" CDX
WITH
1/2" OVERLAYMENT

PART EXISTING ROOF
COVER WITH
NEW FELT AND
SHINGLES

INSULATION
6"-R-19

2-2x6s
PLATE

NO FRAMING
TOUCHES
CHIMNEY
BRICKWORK

NEW
2x6 STUDS
EXTERIOR
WALLS

PART EXISTING ROOF
FRAMING

SKYLIGHT
FRAMING

2x8's x 7'-0"
CANTI-
LEVERED
BALCONY
JOISTS

POCKET
DOOR
FRAME

EXISTING
LOWER FLOOR

FRAMING DETAIL
OF RAISED ROOF
WITH
SUPPORTING WALLS

SKYLIGHT

EXISTING WALL

CLO.

EXISTING
STAIRS

DN

WINDOW
SEAT

CLO

RIDGE

VANITY

NEW
WINDOW

SHELVES

SHELVES

LAVATORIES

SHOWER

SKYLIGHT

2'-8"
POCKET
DOOR

BATH
ROOM

LINEN
CLOSET

NEW WALL
2x6 STUDS

BED

GLASS
DOOR

TUB

PLATFORM

NEW
WALLS
2x6
STUDS

WINDOW

BALCONY
2x6 DECKING

WINDOW

FLOOR PLAN

LEGEND

☐ - EXISTING WALL
■ - NEW WALL

EXISTING
CHIMNEY

SMALL
ROOF VENTS
ABOVE
UNVENTED
RAFTER SPACES

2x10 RAFTERS
ON 16" CENTERS
COVERED WITH
1/2" CDX SHEATHING
15 LB. FELT AND
ROLLED ROOFING

NEW RIDGE VENT
(FULL LENGTH OF ROOF)

CHIMNEY
FLASHING

NEW
SKYLIGHT

EXISTING
2x10 RAFTERS
AND SHEATHING

3'-6"

NEW
SKYLIGHT

ROOF
INSULATION
6"-R-30

VENT

NEW
2x10
RIDGE

2'-8"
POCKET
DOOR

8'

BALCONY
RAILING

NEW FELT
AND SHINGLES

EXIST.
WALL

EXIST. 2 x 10
CEILING JOISTS

2x8's 7'LONG
CANTILEVERED
BALCONY JOISTS
NAIL TO
EXIST. 2x10 JOISTS

INSULATION
6"-R-19

4'

3'-1 1/2"

CROSS SECTION

rity of their home's architectural style, while adding contemporary living space. Equally satisfying, they kept the project within the budget they had set.

Finished in light tones and natural materials—light gray quarry tile, oak trim, white walls, and light gray carpeting—and with white furnishings, the room is a spectacular change from the old pink attic—*Cathy Howard.* *Photos by Karlis Grants.*

PRODUCTS USED IN THIS REMODELING

Fixtures: Steeping bath whirlpool, toilet (Wellworth), Castelle lavs (Country Gray), polished chrome faucets, Kohler Co., Kohler WI 53044. *Tile:* Rustic II (Graystone), U.S. Ceramic Tile Co., 1375 Raff Road, S.W., Canton, OH 44711. *Towels:* Dundee Mills, 111 W. 40th Street, N.Y., NY 10018. *Mirrors:* Hoyne Industries, East Tower, Suite 825, Golf Road, Rolling Meadows, IL 60008. *Heat/fan/ light,switch-plate accessories, concealed paper holder:* NuTone Division, Scovill, Madison and Red Bank Rds., Cincinnati, OH 45227. *Windows, sliding door:* Weather shield Mfg., Medford, WI 54451. *Blinds:* Shadow Gray (bedroom); Garnet Red (bath), Kirsch Co., 309 N. Prospect Street, Sturgis, MI 49091. *Roof window:* Model GGL, Velux-America, Inc., 74 Cummings Park, Woburn, MA 01801. *Track lighting:* Halo/Lighting Products Division, McGraw-Edison, 400 Busse Road, Elk Grove Village, IL 60007. *Furniture:* Techline, Marshall Erdman and Assoc., Madison, WI 53705. *Closet storage system:* Swedish Wire Products, Elfa Division, 1755 Wilwat Drive, Suite A, P.O. Box 861, Norcross, GA 30091.

The backside of the old roof is removed as far as the peak. Temporary braces are secured to front-roof rafters before removal of mating back-roof rafters.

Plumbing vent pipes (below left) will be rerouted horizontally between floor joists to outside wall. Balcony joists (below right) are secured to existing floor joists.

Above, entire back-wall frame is raised into place after assembly on new floor. Roof sheathing is applied, below left. Below right, sturdy scaffolding allows transport of heavy items to the second floor without need to enter the first floor.

Above, the concealed chimney serves as the hub of the new living area. A skylight brightens the bedroom, and a glass door opens onto the balcony. Below left, a trapezoidal window follows the slope of the ceiling. The window seat in front is tile-topped. Below, track lighting on a dimmer switch allows spotlighting effects.

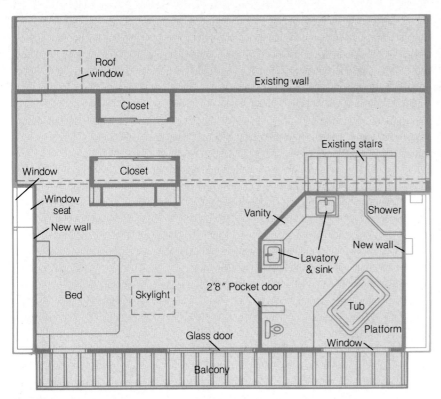

Roof
window

Existing wall

Closet

Window

Closet

Existing stairs

Window
seat

New wall

Vanity

Shower

Lavatory
& sink

New wall

Bed

Skylight

2'8" Pocket door

Tub

Platform

Glass door

Window

Balcony

Floor plan, above, shows positions of
closets, right, with wire basket storage.
The luxurious master bath (below and below right) has dual lavatories. Tongue-
and-groove cedar paneling highlights the
end wall, tub platform, and light soffit.

employing a kitchen designer

Bob and Susan Woods have three children and live in a $125,000 home in Hartford, Connecticut. They knew nothing about kitchens except that theirs didn't work. Clean and neat—yes. But functionally, it was a disaster. What to do about it?

Because both Bob and Susan are busy lawyers, they elected to turn the problem over to a kitchen professional, M. A. Peterson, Inc., of suburban West Hartford. Even before they had to lay out the first dollar, they got a surprising amount of time, plenty of consultation, and a lot of new ideas. Before it was all over, they also got flawless installation of a kitchen they love.

"We knew what we wanted, at least in concept, but could not take the time to do all the contracting ourselves," says Susan. "What we needed was one company or person who would take care of everything. So we called a friend who knows everything and everybody, and where to go for anything." The friend gave them the names of three people, including Ken Peterson, CKD (Certified Kitchen Designer), president of M. A. Peterson. Says Susan: "We called all three and asked them to come see us. But Ken asked us to visit his showroom first, and we did."

The story from that point on is typical of how a kitchen pro works with clients, combining their desires, needs, and ability to pay with a pro's talent and product knowledge.

It was on June 20 that Bob and Susan kept the appointment to visit Ken Peterson's showroom. The visit lasted over an hour. They inspected several displays, saw some ideas and products they liked, and, perhaps even more important, saw some things they didn't want in their kitchen. Impressed with the complete, tasteful displays, the Woods decided they wanted a designer to come to their home. The design associate who had conducted the showroom tour then filled out a "lead card"—a card listing the customer's name, address, and the

Professional kitchen designers can usually introduce you to a wide range of options you might otherwise have overlooked.

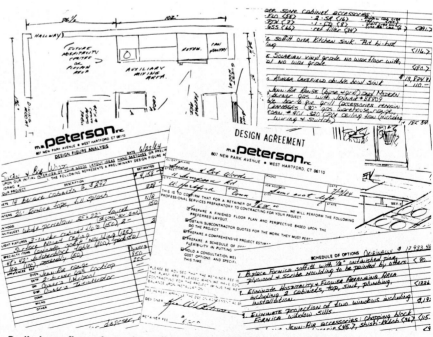

Preliminary floor plans, three-dimensional sketches, and detailed cost estimates let you pare down to your budget before you go to contract.

The result is usually an attractive, step-saving work place that you had a strong voice in designing.

like—and an appointment was made.

Three days later, Ken Peterson himself visited the home. He listened to the Woods describe both their kitchen problems and their remodeling ideas. Ken also asked questions, filling out his "Kitchen Project Profile" form as the conversation progressed. This gave him planning information on the family, the house, traffic patterns, social habits, and kitchen activities.

After about an hour, Ken was ready to describe his ideas to Bob and Susan. Basically, the redesign would involve moving a guest-bedroom door to permit centralizing the kitchen toward that wall, leaving space for a desired dining area at the other end. To help Bob and Susan comprehend, he made freehand drawings of his concept as he discussed it, including a rough floor plan and two perspective sketches.

The ideas made sense to Bob and Susan, so Ken filled in a "Design Figure Analysis" form to give them rough idea of the cost. It came to $13,500. The Woods had expected to hold costs to around $10,000, and wanted time to think about it. Ken's next question: "Are you comfortable enough with the ideas and cost range for me to get exact measurements?" They were, and Ken measured. Total time of his visit: a little over three hours.

For the next three weeks there was no communication. In that time Susan and Bob checked out the two other dealers. One, a millwork firm, offered to install cabinets for $5000, but said the separate dining area was impossible. The other, a kitchen specialist,

said their "his-and-her cooking areas" idea was impossible. So they decided to go with Peterson and they worked out their own financing of the project.

On July 13, Bob and Susan returned to the Peterson showroom to discuss some new ideas and to sign a "Design Agreement"—a one-page contract under which the clients pay a retainer based on 5 percent of the estimated price, in this case $620. They are not yet contracting to buy a kitchen, but they will get a finished floor plan and perspective drawings, subcontractor quotes, a comprehensive estimate, and a final consultation, at which time they decide whether it's go or no-go. The Design Agreement also gives advance notice of payment terms—50 percent on signing a contract (minus the retainer fee), 40 percent on cabinet delivery, the balance on completion.

It was August 6, seven weeks after the initial contact, that Bob and Susan contracted to buy the new kitchen. But in the time between signing the Design Agreement and final contract, Ken's firm did another twelve hours of work—preparing the mathematical layout, subcontractor quotes, final floor plan and perspectives, estimates, and other documents. All were presented at the final consultation. With changes the Woods had requested or approved, costs had by now climbed to $17,933.

"That," says Susan, "was when we really appreciated Ken's professionalism. He had prepared a list of the options, with prices, and we went down the list with him, scratching

items to shave the price. For example, we decided to postpone a hospitality and flower-arranging area we had added. It included two cabinets, a sink, a countertop, and plumbing and came to over $1300. And we had asked for a bay window by the table, but Ken pointed out we already had plenty of light; eliminating the bay saved another $1192. We just went down the list that way and got it back down to $13,874."

The project called for Bellaire custom cabinets, which meant waiting because custom cabinets are not started by the factory until the consumer signs a contract. Eight weeks later the cabinets were delivered. The Woods' kitchen was left intact through this waiting period, as such jobs are never started until all materials are in hand. While Ken scheduled three weeks for the work, it was finished in just sixteen days, on October 19.

What do Bob and Susan think of it now? "It really was fun," answers Bob, "to be involved in the design process. And it was fun to come home each evening and see how much had been done. The butcherblock table gives us a nice place to work as well as eat, and we got a lot more kitchen than we expected."

Susan concurs: "Several of our friends have asked how many feet we added to the room. I'm proud of it, and I think it proves how worthwhile it is to go to a kitchen professional. Ken even loaned us a microwave so we could cook in the dining room while installation was going on—*Patrick J. Galvin.*

host's wine cellar

What was once a catchall for little-used items can become a place to get away to—a tasting room for sharing your prize wines with friends or for hosting intimate wine-and-cheese parties. Line the walls with bottle bins and you've got storage that lets you buy economically now, taking advantage of case discounts. At last you'll have space for "laying down" young wines to let them mature into the fine vintages you'd otherwise have to pay a fortune for in two to ten years.

This is what Syd Dunton, in cooperation with the California Redwood Association (1 Lombard Street, San Francisco, CA 94111), created from a small basement that was previously the location of a gas furnace, water heater, water softener, and a jumble of household paraphernalia.

The basement was a designer's nightmare. It had exposed floor joists overhead, two ugly columns supporting a structural beam, a concrete floor, and a 2-foot-high concrete abutment running around the perimeter, with open stud walls above it.

The job started with insulation: R-11 fiberglass in the stud cavities and a 3-mil polyvinyl moisture barrier stapled to the existing studs. That is covered with 1×12 redwood boards, applied at a 45-degree angle.

The abutment was turned into a bench in some areas, covered by 1×12 redwood planks. In the remaining area it's faced with redwood, to produce a wainscoting effect for the wine bins (seventy-two in all) built above it.

The furnace, along with the water heater and softener, is enclosed behind a partition wall insulated with fiberglass batts. Access to the utilities is through a hinged panel.

The concrete floor is covered with 12×12-inch unglazed terra-cotta tiles mortared in place. The floor joists overhead are covered with tar paper, over which strips of lath are nailed for a trellised effect. Lighting—aside from candles—comes from incandescents in wall sconces. Lights are controlled by dimmer switches.

The only natural light reaching the cellar is through a stained-glass door, which filters the light for pleasant effect and better wine-storage conditions (the less ultraviolet light the better).

Perhaps the cleverest idea is the treatment of the support columns. One is built into a storage divider. The other becomes the support for a small round table where wine and cheese are served.

To add the look of age appropriate to a wine cellar, decorative corbels (cut from redwood) trim the beams, and the redwood throughout is distressed and mellowed with an oil stain.

The besement stays cool, a condition essential for proper storage of wine, thanks to the isolated heating devices. The utility enclosure also helps keep vibrations away from the wine bottles, which store better in peace and quiet—*Charles A. Miller.*

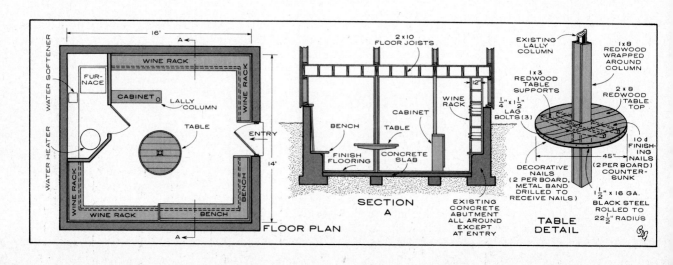

creating a basement suite

A home is made for enjoyment, but if your family is typical, full enjoyment calls for more room than you've now got. If your house has a basement that's currently being used only for storage (or housing the family pet), additional living space is readily at hand.

The basement of the town house shown here was a combination laundry room–storage dump before being converted into this attractive family suite. It was created by Masonite Corp. as a showcase for prefinished hardboard paneling. The panels shown are called Woodfield, a narrow-groove textured board that comes in standard 4 × 8-foot sheets.

To achieve a similar transformation of your basement, start with a plan to scale (½ inch = 1 foot) that shows the location of all permanent objects such as the furnace or boiler, tanks, washtubs, support posts, and so forth. Try to isolate utility items from the living area by carefully plotting the location of partition walls. In our example (following), a stud wall of 2 × 4s on 16-inch centers was erected between the laundry area and the rest of the basement.

Next, in red pen, draw in the location of lights, vents, exhaust fans, and other obstacles that extend from the ceiling. Now draw a pattern of 1-inch squares (representing 2 × 2-foot ceiling tile) on a piece of tracing paper. Move the tracing paper over the plan so that the obstacles clear any of the seams in the tile. You may find that you'll have to reposition lights and other movable fixtures to ensure that support strips don't interfere (given the limitation of a reasonable starting point for the tile).

The Woodfield paneling from Masonite is designed to be butted together over stud framing. Use adhesive or color-coordinated finishing nails vertically spaced 12 inches apart to mount the panels directly to the studs. Paint the edge of all studs that stand behind panel joints black so that if the panels contract later because of weather, the gap won't be noticeable.

Use 2 × 2 stock to fur out along perimeter walls except where obstructions must be cleared. Where small pipes and conduit protrude, frame out the entire wall length with 2 × 4s.

Match these "before" shots with their color counterparts to appreciate what partitions, paneling, and furnishings can achieve. Study/entertainment area at left began as waste space in top photo. Recessed shelving in cozy dining nook is built into partition that hides old laundry area ("before" shot above). Open stairwell (below) was enclosed for storage.

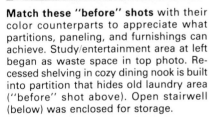

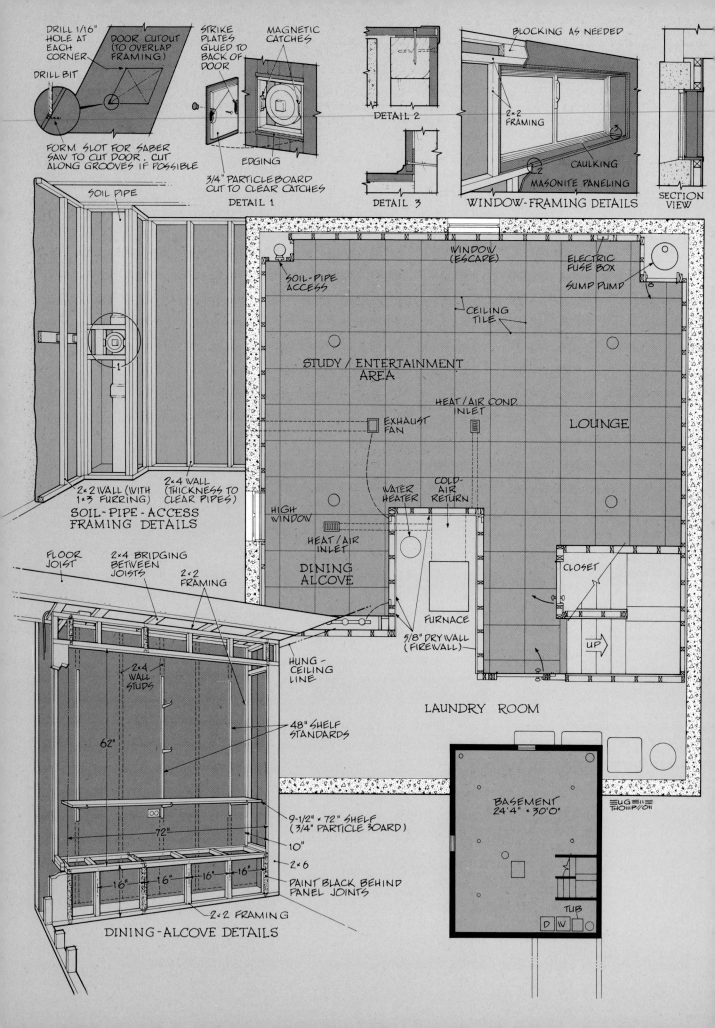

DRILL 1/16"
HOLE AT
EACH
CORNER

DOOR CUTOUT
(TO OVERLAP
FRAMING)

STRIKE
PLATES
GLUED TO
BACK OF
DOOR

MAGNETIC
CATCHES

DRILL BIT

FORM SLOT FOR SABER
SAW TO CUT DOOR. CUT
ALONG GROOVES IF POSSIBLE

EDGING

3/4" PARTICLEBOARD
CUT TO CLEAR CATCHES

DETAIL 1

DETAIL 2

DETAIL 3

BLOCKING AS NEEDED

2×2
FRAMING

CAULKING

MASONITE PANELING

WINDOW-FRAMING DETAILS

SECTION
VIEW

SOIL PIPE

2×2 WALL (WITH
1×3 FURRING)

2×4 WALL
(THICKNESS TO
CLEAR PIPES)

SOIL-PIPE-ACCESS
FRAMING DETAILS

WINDOW
(ESCAPE)

ELECTRIC
FUSE BOX

SUMP PUMP

SOIL-PIPE
ACCESS

CEILING
TILE

STUDY / ENTERTAINMENT
AREA

LOUNGE

HEAT/AIR COND.
INLET

EXHAUST
FAN

HIGH
WINDOW

HEAT/AIR
INLET

WATER
HEATER

COLD-
AIR
RETURN

DINING
ALCOVE

CLOSET

UP

FURNACE

5/8" DRYWALL
(FIREWALL)

LAUNDRY ROOM

FLOOR
JOIST

2×4 BRIDGING
BETWEEN
JOISTS

2×2
FRAMING

2×4
WALL
STUDS

62"

48" SHELF
STANDARDS

HUNG-
CEILING
LINE

72"

9-1/2" × 72" SHELF
(3/4" PARTICLE BOARD)

10"

2×6

PAINT BLACK BEHIND
PANEL JOINTS

16"   16"   16"   16"

2×2 FRAMING

DINING-ALCOVE DETAILS

BASEMENT
24'4" × 30'0"

TUB

D   W

OP PLATE NAILED TO JOISTS OR TRUSSES

PAINT STUDS BLACK BEHIND PANEL JOINTS

PLUMB WITH LEVEL AND SHIM BEFORE NAILING TO 3 FURRING

SHIMS

3 FURRING ATTACHED TO FOUNDATION OVER VAPOR BARRIER

2×2 FRAMING

VAPOR BARRIER

16" 16" 16"

2×2 BASE PLATE

VAPOR BARRIER

DETAIL 4

2×2 FRAMING

2-1/2" CHALK LINE

2×2 WALL-FRAMING DETAILS

ALIGN NEW PANELING FLUSH WITH EXISTING DRYWALL

MASONITE PANELING

EXISTING POST

DOOR

MASONITE PANELING

EXISTING WALL

JAMB

STAIR STRINGER

1/4 ROUND

NEW 2×4 FRAMING

CHALK LINES

USE WOOD SHINGLES AS WEDGES BETWEEN FRAMING AND JAMBS

CORNER BEAD

DETAIL 5

CLOSET AND STAIRWELL DETAILS

Frame, too, around soil stacks, valves, electrical panels, and other fixtures that you'll have to get to from time to time so that access can be easily provided.

In our example, a small door was built in the framework around a soil pipe for access to the cleanout. A corner was built out and a full-length door added to maintain a passage to a sump pump.

To build small access ways, first measure the obstruction to determine the best location for an opening. Now, cut the framing for this opening to size. With a sheet of paneling lying face up on the floor, position your framing for the neatest effect with the panel grooves, letting the cut lines fall on the vertical grooves wherever possible.

Now drill 1/16-inch holes at each corner of the opening. Mark a saw line between the holes. At two opposite corners drill several more holes close together along the saw line, forcing the drill back and forth to make a slot for the saber-saw blade.

Now turn the panel over and mark the cutting line of the opening between the corner holes, slip the saw blade into the drilled slots, and cut out the access door.

When you're finished, sand the edges of the opening smooth. If you put edging strips around the opening, any sawing imperfections will be concealed.

Glue a piece of particleboard or plywood to the back of the piece you've cut out to make a cover, and add a small handle and strike plates. Install the framework and add magnetic or other suitable catches to hold the cover in place.

Our stairwell closet was framed with 2×4s and covered with 1/2-inch drywall before mounting the paneling. The door framing was positioned so that the door was centered on the short wall. Measure the distance from the edge of the panel to where the door opening is to be cut out, and adjust so that at least one cut line will fall along a panel groove.

Lay the door panel in position on the back of the paneling and trace its outline. Remove the door, put the panel on saw horses, and carefully cut the outline (from the panel's back) using a saber saw or trim saw, being sure to pad the front so as not to scar the finish.

Coat both the door and the back of the cutout paneling with contact cement and bring the coated surfaces together when the cement is tacky. (You'll want to use the slipsheet technique to ensure alignment, since you can't separate the pieces once you've started.) Roll the paneling with a rolling pin to be sure the bond is tight around all edges.

Now nail the wall paneling in place and hang the door. Add the stairwell paneling, including the corner beads. Note that the lower edge of the panel is cut on an angle to match the top of the stair stringer. Any imperfections along the cut edges of the paneling can be hidden by trim pieces.

The bookshelves are framed with 2×4 and 2×2 stock. After the main dividing wall is framed, install the back sheet of paneling before building the front section of the alcove. Install electrical outlets, switches, and overhead lights. The apron at the top is ideal for a 48-inch fluorescent fixture. Build the framework, then cut the rest of the paneling to size and install throughout, being sure to use inside and outside corner molding where panels meet at 90 degrees.

To install the ceiling, the lip of the perimeter support-angle pieces should be at least 3 inches below the lowest duct or pipe to allow enough room for you to insert the tile after the support grid is up. However, even if there are no projections, you should allow 6 inches of free space below the joists.

To show the position of the perimeter support-angle pieces, snap a chalk line (with the aid of a helper) around the room at a height that will allow inserting the tiles without interference. Use a level to be sure the line is perfectly horizontal all around.

Now install the support-angle pieces. Plan to run the main hangers of the system at right angles to the joists so as to avoid having to bridge between them. Install the hangers and tile.

In our example, the ductwork for the heating and cooling system of the house was extended and vented into the new family room. Air conditioning helps keep the new addition dry. In addition, an exhaust fan was installed to keep moisture and temperature-related expansion and contraction of the paneling to a minimum.

Once you've installed your flooring, you will be ready to move in your furnishings—and start living more spaciously—*Charles A. Miller*.

The materials for this project were supplied by these companies: *Paneling:* Woodfield Design paneling, Masonite Corp., 29 N. Wacker Drive, Chicago, IL 60606. *Furniture:* Marin County Collection upholstery and occasional tables, Composite Collection party table and chairs, desk, Riverside Furniture Corp., Fort Smith, AR 72902. *Ceiling:* Cumberland Pattern acoustical ceiling, Armstrong World Industries, Lancaster, PA 17604. *Carpeting:* Form III, North Vernon, IN 47265.

daylighting your basement

A sun-bathed basement wall is what prompted this homeowner to install a bay window to create a cheery bedroom for two small girls. Even if your house sits a lot lower in the ground, you can do the same by regrading or digging a window-well.

Breaking through the wall takes a cold chisel and hammer for block; a power chisel or rented jackhammer for poured concrete. (Wear safety goggles.) Smooth the opening's edges and butter them with a stiff mortar mix. Then set in a three-sided frame of nominal 2-inch-thick lumber as wide as the concrete is thick. The existing house plate serves as the top of the frame. Plumb and level the frame while the mortar is still wet—this is the rough opening for the window you will install.

The bay window should be built in advance (see illustration), so you can make the frame and opening to match. The stock window need not be double-hung, of course; a casement window would be easier to open at arm's length, for example. Assembled ready-to-install bay and bow windows are available in standard sizes. You may want to consider these before making the rough opening.

Double-hung window, flanked by fixed sash, forms bay unit (left). Set into opening cut through foundation wall, it's shingled on top. Roof could also be copper-covered plywood. Besides light and air, deep-set window provides wide shelf or seat (below). Narrow cedar trim frames window, matches wood ceiling. Valance hides the window shade; heat register is above it.

Photos: Western Wood Products Assn.

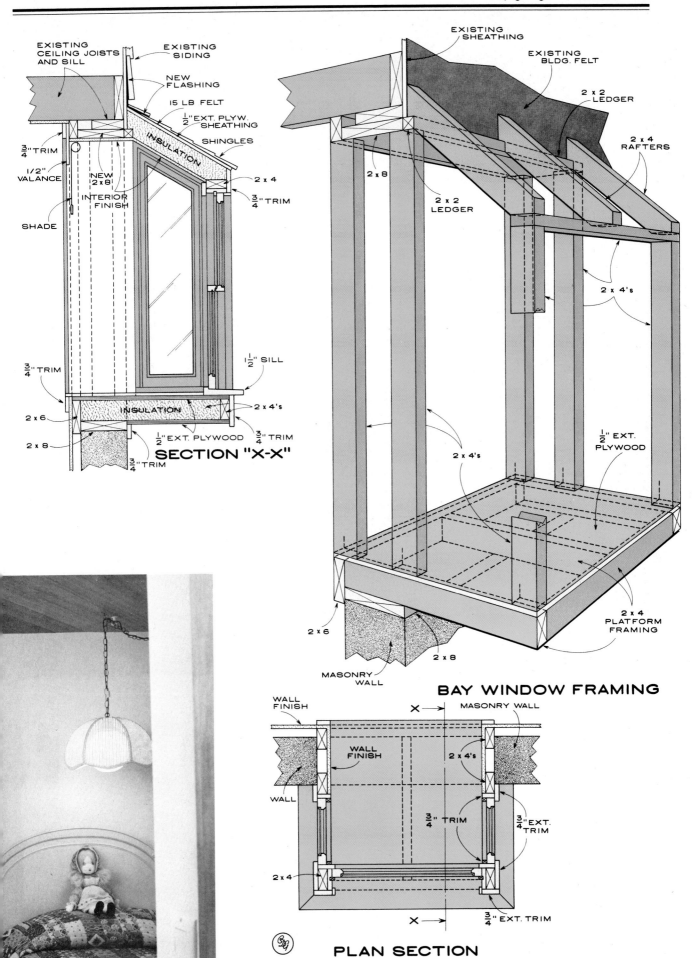

EXISTING
CEILING JOISTS
AND SILL

EXISTING
SIDING

NEW
FLASHING

15 LB FELT

$\frac{1}{2}$" EXT. PLYW.
SHEATHING

SHINGLES

$\frac{3}{4}$" TRIM

1/2"
VALANCE

NEW
2 x 8

INTERIOR
FINISH

INSULATION

SHADE

2 x 4

$\frac{3}{4}$" TRIM

$\frac{3}{4}$" TRIM

$1\frac{1}{2}$" SILL

INSULATION

2 x 6

2 x 4's

$\frac{1}{2}$" EXT. PLYWOOD

$\frac{3}{4}$" TRIM

2 x 8

$\frac{3}{4}$" TRIM

SECTION "X-X"

EXISTING
SHEATHING

EXISTING
BLDG. FELT

2 x 2
LEDGER

2 x 4
RAFTERS

2 x 8

2 x 2
LEDGER

2 x 4's

2 x 4's

$\frac{1}{2}$" EXT.
PLYWOOD

2 x 4
PLATFORM
FRAMING

2 x 6

2 x 8

MASONRY
WALL

BAY WINDOW FRAMING

WALL
FINISH

MASONRY WALL

X

WALL
FINISH

2 x 4's

WALL

$\frac{3}{4}$" TRIM

$\frac{3}{4}$" EXT.
TRIM

2 x 4

X

$\frac{3}{4}$" EXT. TRIM

PLAN SECTION

adding three feet to a garage

What do you do when you buy a second car and your garage is meant for only one?

When I faced the problem, my friends offered ideas galore. One advised separating the garage at the peak and moving the walls over. Another suggested tearing the whole thing down and starting from scratch.

My own solution was cheaper and more practical. Since a mere 3 feet more space would accommodate the new car, I decided to detach one wall and move it out.

I started by taking off the old siding—but very carefully; I wanted to reuse it, both to cut down on costs and because it matches the house. I also reused as much other material as possible—insulation board, 2×4s, 2×6s. The old garage door, which I couldn't reuse, I sold for $70. Since I didn't like the existing window, I purchased a new one for $50.

I poured the foundation for the new wall 3 feet deep and 12 inches wide.

Extending the roof 3 feet was simply a matter of splicing a piece to the ridge board and adding plates, studs, and three rafters on each side. I was able to find shingles to match the original

Front of the garage is shown above during the construction period. Two 2×8s were used over the door opening to support the new double doors. Old siding was reused to save money.

Rear view of the garage extension can be seen in photo above as portion of the roof that is not yet shingled. Shown at right is the garage interior, with side wall finished off and new window in place.

ones. I hired a commercial firm to hang the door; it cost $250.

While measuring the floor, I discovered that one corner was a good 4 inches below level, creating a small water basin. I knew that just applying another thin layer of concrete would be impractical—sooner or later, it would break up—and I didn't want to pour a whole new floor. Instead, I had the floor blacktopped and then painted it light gray.

Outside, I built a small brick wall to the left of the garage entrance and installed a wrought iron post similar to one at the entrance of the house. The result: A two-car garage that looks like it was always there—*Richard C. Redmond.*

Finished two-car garage, with decorative wall and post to match house.

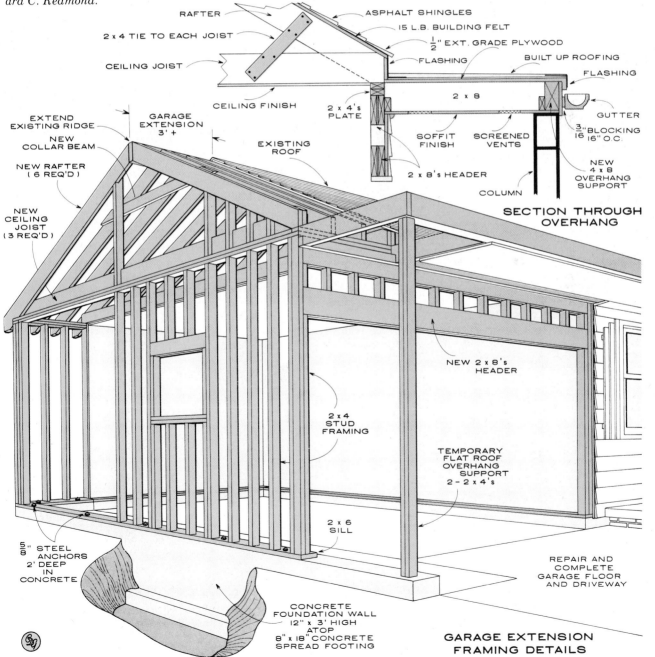

RAFTER

2 x 4 TIE TO EACH JOIST

CEILING JOIST

CEILING FINISH

ASPHALT SHINGLES

15 L.B. BUILDING FELT

½" EXT. GRADE PLYWOOD

FLASHING

BUILT UP ROOFING

FLASHING

2 x 8

2 x 4's PLATE

SOFFIT FINISH

SCREENED VENTS

GUTTER

3/16" BLOCKING 16" O.C.

2 x 8's HEADER

COLUMN

NEW 4 x 8 OVERHANG SUPPORT

SECTION THROUGH OVERHANG

EXTEND EXISTING RIDGE

NEW COLLAR BEAM

NEW RAFTER (6 REQ'D)

NEW CEILING JOIST (3 REQ'D)

GARAGE EXTENSION 3' +

EXISTING ROOF

NEW 2 x 8's HEADER

2 x 4 STUD FRAMING

TEMPORARY FLAT ROOF OVERHANG SUPPORT 2-2 x 4's

2 x 6 SILL

5/8" STEEL ANCHORS 2' DEEP IN CONCRETE

CONCRETE FOUNDATION WALL 12" x 3' HIGH ATOP 8" x 18" CONCRETE SPREAD FOOTING

REPAIR AND COMPLETE GARAGE FLOOR AND DRIVEWAY

GARAGE EXTENSION FRAMING DETAILS

guide to windmills

D own the road, still at some distance, a rhythmic, undefined flashing caught my eye. Drawing closer, I made out a tall, white tower topped by an aerodynamically trim pod. At one end of the pod, graceful blades seized the sun and sent back their almost hypnotic flick, flick, flick.

I was watching a modern wind generator at work. Interestingly, the base of the windmill's tower was firmly anchored in the property of the local power company.

Perhaps I really was a bit hypnotized by the blades flashing in the sun. As I drove on, my mind harked back to a summer afternoon ten years ago. As I watched, a ragtag group of young wind-power enthusiasts, collectively calling themselves Windworks, stood shouting advice to each other and tugging on guy ropes while one of their number hung precariously on a rickety ladder as he mounted the blades of an experimental hand-built wind turbine. But that was ten years ago. Now, the sometimes crude experimental rigs have evolved into sleekly efficient machines such as the latest Windworks 10-k W generator in the power company's yard.

Today more than thirty-three wind generators and variations of them are being sold by at least twenty-five different manufacturers. If you have the wind, the place, and the need, there is almost certainly a wind generator you can buy and put to work.

Over two dozen wind-turbine builders were queried on the sizes, capacities, and details of their machines. Wind-energy pioneer Marcellus Jacobs of Jacobs Wind Electric Co. and Hans Meyer of Windworks were also consulted. The consensus: Despite the many designs to choose from (and which we'll talk about), making a decision comes down to the question of how you want to use wind power. Do you want to be totally independent of the power company? Or do you want

to augment the utility's supply, thereby cutting your electric bill?

When you start dreaming of wind power for your home or vacation cabin, be mindful of these two basic routes to follow: You can choose a modest machine producing from ½ kW to about 3 kW of direct current to charge a bank of batteries or to feed resistance heaters; or you can select a larger, higher-output AC machine of 5 kW or more with a powerline tie-in—called an inter-tie—to control the electrical output and provide a backup power source.

Feeding batteries with a small wind generator, such as the Altos BPW-12A or Dunlite 2000, allows you to accumulate power during the times you don't need it and draw on that accumulation for surge and steady loads. The line inter-tie does the same job differently. Using a line inter-tie is somewhat like putting electricity in the power company's energy bank. Power you produce but don't need you deposit, in a sense, by feeding it into the utility lines. When you need it you take it back. The inter-tie does away with the batteries. Product Development Institute's WJ6500 is an example of an inter-tie machine, as is Enertech's 4000.

Of course, direct current straight from the wind generator can be used. In a heating application the Bergey 1000 series or Whirlwind's 3000 might do just fine and simplify your installation by eliminating batteries. The Aero-therm Wind Furnace is designed only to power resistance heaters—the largest machine devoted to that purpose.

However, the power-line inter-tie, in addition to providing a backup when you need more power than the wind can produce, gives you other advantages. For instance, the output from your wind system is held to exactly the same alternating frequency and wave characteristics as utility power. And in most areas you can profit from the

excess power by selling it to the utility at a rate established by your state.

The perfect answer?

There are several ways to inter-tie, but one method now becoming quite popular is using the induction-type generator. (Note that nine of the wind-generator makers in the table at the end of the article offer induction units.) An induction generator is actually much like the induction motors in your home, but when it's driven by the wind it becomes a generator.

The magnetizing power—called excitation—for an induction generator comes from the utility line. Your generator output, by the nature of electrical power generation, automatically matches the frequency and wave form of the utility power.

Another extremely important advantage of the induction setup, at least from the standpoint of a lineman repairing a downed line, is that as soon as the line power fails, your generator goes dead. There's no way you can feed back potentially lethal power to the line. By the same token, of course, your windmill stops supplying you with power.

Though the induction generator sounds like a perfect answer, many generator makers, including Marcellus Jacobs, disagree. Why? Because induction generators are locked into a fixed rotational speed by the power line. Says Jacobs: "Fixed-speed propellers [on induction generators] are usually running too fast or too slow . . . and do not fully harness the wind energy available." Jacobs's own AC machine delivers 240 V through an inverter system, which allows the blades to vary in speed with the wind.

A few years ago wind-power experimenters were in a fever of invention, with their strange and wonderful turbines shaped and rigged in every imaginable fashion. Their goal was to extract every fraction of power from the wind stream.

The exotic windmills have largely been abandoned, though there are still a few fascinating designs with striking differences in structure, blade materials, blade-pitch controls, overspeed controls, generator drives, and generators—all of which influence durability. (And durability under severe weather conditions is probably the single most important quality of any wind generator.)

The Windstar WS-462-16 with seven pairs of blades, the Altos with a bicycle-wheel rotor, the Pinson Cycloturbine, and the Tumac 680-1 (Darrieus) are refinements of machines that seemed novel during those visionary times. The Tumac and Pinson machines are among the few vertical-axis windmills aimed at the home market.

The Pinson Cycloturbine has the advantage of accepting wind from any direction without orienting to it; it's also self-starting. The Tumac Darrieus is also omnidirectional but requires using the generator as a starter motor to bring the rotor up to speed when a sensor indicates the wind is adequate to produce power.

Upwind or downwind positioning on the more conventional horizontal-axis

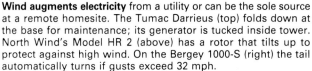

Wind augments electricity from a utility or can be the sole source at a remote homesite. The Tumac Darrieus (top) folds down at the base for maintenance; its generator is tucked inside tower. North Wind's Model HR 2 (above) has a rotor that tilts up to protect against high wind. On the Bergey 1000-S (right) the tail automatically turns if gusts exceed 32 mph.

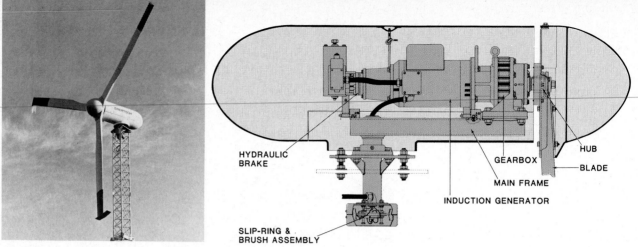

Downwind turbine from Enertech has a hydraulic brake that locks the rotor if the power line fails. By design, its induction generator shuts down immediately upon failure so the machine can't backfeed the utility and endanger linemen.

machines is still a hot and unresolved controversy. Upwind rotors are steered into the wind by a tail vane. Advocates of the upwind system point out that they avoid the wind shadow of the tower for an even flow of wind. The new upwind 10-kW Jacobs even cants the turbine and power head slightly upward to meet the downward pitch of the wind caused by ground effect.

At the other end of the spectrum is the downwind design, such as the new Windworks generator. What's the advantage of each type? On smaller units, says Hans Meyer of Windworks, you can use a tail vane both for positioning and for other control schemes, such as overspeed protection. As rotor diameters increase, however, "a tail structure that puts the tail sufficiently clear of the slipstream becomes a big and expensive structure."

There's also considerable disagreement about blade materials. The best argument for wooden blades is probably their long and successful history of use. Jacobs points out that the spruce-blade generators installed at Byrd's Little America camp in the Antarctic in 1933 were still running in 1955 after years of neglect. Nevertheless, aluminum, in cast, formed-sheet, and solid shapes, is used in nearly half of the windmills in our table. Fiberglass is also used extensively. Windworks uses both materials. Windstar uses stainless steel.

The really intriguing design differences among wind-machine manufacturers involve blade-pitch control and overspeed protection. Turbines with small rotors often use no pitch control or rely on flexing blades to conform to variations in wind speeds. Larger machines, however, require some form of governing to adjust the blade-pitch angle for both efficiency and speed control. Nearly all use centrifugal force as a sensing agent, although

some monitor electrical output since it relates to speed. Others, such as Bergey, mount strategically placed weights on the blades to twist them into different pitches in response to speed changes. Weights behind the blades or in the power head can also be used to actuate blade-change linkage or a hydraulic system that in turn moves the blades.

There is simply no agreement on pitch controls. As a rule of thumb, however, the more sophisticated the pitch control, the more complex the mechanism. As Karl Bergey of Bergey Windpower points out, "High aerodynamic efficiencies are obtained with . . . variable-pitch rotor systems at the cost of mechanical complexity and decreased reliability."

If there is a middle ground it's probably the Jacobs blade-actuated system, which floats the blades on strong springs at the hub and uses the force of the blades themselves to power the pitch-change movement without electrical or centrifugal sensors or tricky mechanical linkages.

But pitch control is not the only challenge involved. The wind varies mightily in both direction and intensity. There are few, if any, locations in the United States where occasional gusts approaching 100 mph or higher are not encountered. Left to spin unhindered in such gusts a rotor tip might speed up from a comfortable 175 mph in a 25-mph wind to 300 to 500 mph. Destruction is almost certain if this occurs.

The simplest overspeed control is to put a brake on the rotor and run a cable to the ground. When high winds threaten, you tug the cable and secure the rotor. It can work, if you're there and if you can get to the tower to pull the cable soon enough.

However, most of the upwind generators have some form of pivoting tail to swing the rotor around (out of direct

wind) at about 30 mph. Some simply tuck the rotor and shut down. Others, such as the Bergey, position the rotor so it keeps turning, but at a reduced power output. Jacobs's machine maintains a reduced power output by both the pivoting action of the power head and a pitch control that turns the blade angle to 60 degrees. Jacobs says this will protect the machine from winds of up to 150 mph.

Without a tail vane, of course, this type of pivoting is not available. Downwind rotors, such as the North Wind L16, use a pitch control to feather the blades.

Key to success: survival

Karl Bergey feels that as wind-generator manufacturers "shake out," only those companies with machines that can endure severe conditions are likely to be survivors. Says Jacobs, who's been in the wind-power business since 1930, "The secret to success in the wind-electric business is quite simple: Survive in the real world of Mother Nature year after year."

Wind generators, although turned by a turbine rotor, behave much the same as the alternator in your car driven by a belt from the engine. Idle at a stop sign in your car and the alternator light may wink on to tell you your alternator is turning too slowly to produce the power needed to carry your lights, air conditioning, and other loads. If the wind drops below about 10 mph—the average cut-in speed for a wind generator—you'll no longer have useful power. In your car, of course, you usually have plenty of engine speed for the alternator (which typically works best at 1800 to 3600 rpm). But on a wind generator, to maintain a safe blade-tip speed, rotors are generally limited to between 600 and 700 rpm.

One solution is to use a gear drive to speed up the generator. Not sur-

Windmills you can buy

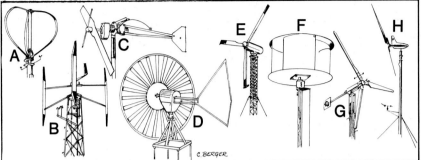

C. BERGER

Maker	Model	Type[1]	Blade dia. (ft.)	Blade material[2]	Blade control	Drive[3]	Generator	Output (V)[4]	Rating (kW)	Use[5]	Price ($)[6]
Aeolian Energy	AEO-20	E	20	FB	Feather	Gear	Induction	110 AC	4	Gen.	9,950 (with tower)
	AEO-32	E	32	FB	Feather	Gear	Induction	220 AC	10	Gen.	17,745
Aero Power Systems	SL-1000	G	10	Wood	Centrifugal	Gear	Alternator	14.5 DC	1	BC	4,000
	SL-1500	G	12	Wood	Same	Gear	Alternator	12/24/48/110 DC 110 AC	1.5	BC	4,200
Aero-therm	Wind Furnace	E	33	FB	Electro-mechanical	Gear	Alternator	480 AC	25	RH	26,995
Altos International	BPW-12A	D	11.5	AL	Turning tail vane	Gear	Alternator	115/200 AC	2.2	BC	3,500 to 5,000
AWI	10-I	G	26	FB	Centrifugal	Gear	Induction	240/480 AC 60Hz	10	Gen.	11,500
Bergey Windpower	1000-S	G	8.3	AL	Weighted flexible blades	Direct	Permanent-magnet alternator	115 AC	1	Gen.	2,995
Bertoia Studio	A.P.S./H	E	20	AL	Centrifugal	Direct	Permanent-magnet alternator	36/48 DC	2	BC	1,500
Dunlite	2000	G	13	AL	Feather	Gear	Alternator	110 DC	2	BC	7,685
Enertech	4000	E	19.6	Wood	Sensor & brake	Gear	Induction	230 AC	4	Gen.	11,000 to 16,000 installed
	1800	E	13	Wood	Sensor & brake	Gear	Induction	115 AC	1.8	Gen.	6,000 to 10,000 installed
Future Energy R&D	Zeus 16	G	16	FB	Feather	Gear	Induction	120/240 AC	3	Gen.	3,975
	Zeus 26	G	26	FB	Feather	Gear	Induction	120/240 AC	10	Gen.	9,475
Jacobs Wind Electric	10-KW	G	23	Wood	Blade governor	Hypoid gear	Brushless alternator	240 AC	10	Gen.	20,000+ installed
Kucharik Wind Electric	FDG63-R	C	5	Wood	Folding tail	Gear	Permanent-magnet alternator	12 DC	0.5	BC	545
Millville Hawaii Windmills	15-1-1IND	C	28.8	AL	Mechanical	Gear	Induction	240 AC	15	Gen.	16,500
Natural Energy Systems Unlimited	WV 50 G	G	16.4	Wood	Feather	Gear	Permanent-magnet alternator	110 AC	6	Gen.	9,500
	WVG 120 G	G	19.6	Wood	Feather	Gear	Same	110 AC	10	Gen.	13,500
North Wind Power	L16	H	30	Wood	Mechanical	Direct	Alternator	240 AC	14	Gen.	25,000
	HR2	G	16	Wood	Tilts up	Direct	Alternator	110 DC	2.2	BC	9,800
Pinson Energy	Cycloturbine C-3	B	18	AL	Cyclic	Gear	Induction	240 AC	5	Gen.	10,000
Product Development Inst.	WJ6500	G	20	FB	Sensor & brake	Gear	Induction	240 AC	8	Gen.	11,785
Sencenbaugh Wind Electric	1000-14	G	12	Wood	Folding tail	Gear	Alternator	12 DC	1	BC	3,600
	500-14HDS	G	6.83	Wood	Folding tail	Direct	Alternator	12 DC	0.5	BC	1,680
Tumac Ind.	680-1	A	19	AL	Brake	Gear	Induction	230 AC	9.2	Gen.	14,500
Whirlwind Power	3000	C	14	Wood	Pilot rotor	Direct	Permanent-magnet alternator	32 DC	3	BC	3,600
Winco	W450	C	8	Wood	Centrifugal governor	Direct	Alternator	12/24 DC	0.45	BC	975
Wind Energy	603	F	6	AL	n.a.	Traction	Induction	110 AC	1	Gen.	2,000
	808	F	8	AL	n.a.	Traction	Induction	110 AC	5	Gen.	5,000
Windstar	WS-462-16	D	16	SS	Steering tail	Direct	Permanent-magnet alternator	220 AC	4	Gen.	9,600
Windworks	Windworker 10	E	33	AL & FB	Hydraulic governor	Direct	Permanent-magnet alternator	240 AC	9	Gen.	29,500
Zephyr Wind Dynamo	Windtilter	C	7.4	Wood	Pivots up	Direct	Permanent-magnet alternator	12 DC	0.5	BC	1,200 to 1,500

[1]A: vertical axis, three blades, Darrieus; B: vertical axis, three blades, Cycloturbine; C: horizontal axis, upwind, two blades; D: horizontal axis, upwind, multi-blade; E: Horizontal axis, downwind, three blades; F: Savonius; G: horizontal axis, upwind, three blades; H: horizontal axis, downwind, two blades.

[2]FB: fiberglass; AL: aluminum; SS: stainless steel.

[3]Type of connection between rotor and generator.

[4]Although outputs (AC or DC) and voltages are given for units of greatest general interest, many makers offer great variations in output and sometimes single- or triple-phase outputs.

[5]Gen.: general household power or like application, although some form of electrical control between generator and house or utility lines may be required; BC: battery-charging system; RH: powers resistance heaters.

[6]Prices given are approximate and are for the rotor-generator only; cost of tower, controls, batteries, inverters, and other accessories must be added except where price is given for installed unit.

prisingly, gears, gear bearings, and gear lubrication have caused more than their share of problems in wind generators. Part of this is probably growing pains on the part of generator makers who have had to use gears and gearboxes not specifically designed to operate in severe temperatures and under highly variable loads. Perhaps the difficulty of checking lubricants at the top of a high tower is also involved. Another factor may be the interplay of frequencies and vibrations stemming from rotor frequency, wind shadow, and the generator's magnetic frequencies.

One of the most interesting solutions to the gear-drive problem is Jacobs's hypoid drive. It runs at right angles to the rotor axis and straight down to the generator, which is mounted vertically in the tower. This puts a very short gearbox dead center on the axis about which the machine pivots, and does away with the generator's weight as a concern in yaw.

Other machines, such as those from North Wind, Whirlwind, Windworks, and Bertoia (large rotors) as well as Bergey and Sencenbaugh (smaller rotors), have elected to use special low-speed alternators with no gears and direct drive.

For an approximation on savings, your generator manufacturer should be able to supply a table of average kilowatt-hours per month (versus average wind speed) to be expected from his machine. You can then add up the kilowatt-hours for a year and apply the total against your expected utility cost to estimate how long it will take your system to pay its way—*E. F. Lindsley.*

LIST OF MANUFACTURERS

Aeolian Energy, Inc., R.D. 4, Ligonier, PA 15658; **Aero Power Systems**, 2398 Fourth Street, Berkeley, CA 94710; **Aero-therm Corp.**, Box 574A, R.R. 1, Lenhartsville, PA 19534; **Altos Corp.**, 5420 Arapahoe, Boulder, CO 80303; **Astral/Wilcon (AWI)**, Box 291, Millbury, MA 01527; **Bergey Windpower Co., Inc.**, 2001 Priestley, Norman, OK 73069; **Bertoia Stuido**, 644 Main Street, Bally, PA 19503; **Dunlite Electrical Products/Enertech Corp.**, Box 420, Norwich, VT 05055; **Future Energy R&D Corp.**, Carretera Estatal No. 113, Cruce Carretera No. 477, Zona Industrial, Quebradillas, PR 00742; **Jacobs Wind Electric Co.**, 2720 Fernbrook Lane, Minneapolis, MN 55441; **Kucharik Wind Electric**, Box 786, Toms River, NJ 08753; **Millville Hawaii Windmills, Inc.**, 3028 Ualena Street, Honolulu, HI 96819; **Natural Energy Systems Unlimited**, Davis Road, Prattsburg, NY 14873; **North Wind Power Co.**, Box 556, Moretown, VT 05660; **Pinson Energy Corp.**, Box 7, Marstons Mills, MA 02648; **Product Development Institute**, 4445 Talmadge Road, Toledo, OH 43623; **Sencenbaugh Wind Electric**, Box 11174, Palo Alto, CA 94306; **Tumac Industries, Inc.**, 650 Ford Street, Colorado Springs, CO 80915; **Whirlwind Power Co.**, 5030 York Street, Denver, CO 80216; **Winco** (Division of Dyna Technology), 7850 Metro Parkway, Minneapolis, MN 55420; **Wind Energy**, 1632 Market Street, Denver, CO 80202; **Windstar Corp.**, Box 151, Walled Lake, MI 48088; **Windworks, Inc.**, Rte. 3, Box 44A, Mukwonago, WI 53149; **Zephyr Wind Dynamo Co.**, Box 241, Brunswick, ME 04011.

low-cost kit
solar heaters

When Henry Ford brought out his mass-produced Model T in 1913 for $500, some 200 other automakers were selling hand-built cars for over $2000.

Jack Hanson sees a parallel between what happened to automobiles in 1913 and what's happening to solar heaters in 1984. His mass-produced system has a $495 price tag while conventional systems start at well over $2000.

Hanson is a self-educated aircraft and missile designer, solar architect, and inventor who likes to "improve on things."

"In 1978," Hanson recalls, "my wife said we really had to do something about the fuel bills. Our old 7000-square-foot building was eating us up. My wife suggested I try solar collectors, but I told her it wouldn't be cost-effective unless I could double the heat

output and quarter the usual collector cost.

"I decided to experiment anyway, since I had a leaky roof that I could cover with a collector instead of repairing. The collector was 75 feet by 16 feet, and my calculations predicted a 25 percent fuel saving. But when we went over the bills later, we found the saving had actually been 65 percent," Hanson enthused.

That started him on a three-step development program. First, Hanson says, he improved heat output by a factor of 5—making every square foot of collector area heat 5 cubic feet of air. This allowed him to shrink the collector to one-fifth its original size.

Savings in materials, fabrication, and handling have made the low price possible. Smaller size and weight—the collectors measure only 37 by 38 inches and weigh 21 pounds—also

simplify installation. Modular design allows you to start with a basic two-panel system and add on later.

Tests show that on average the collectors raise air temperature about 30 degrees F. A pair will comfortably heat a $10 \times 12 \times 8$-foot room when it's 20 degrees outside.

"This kind of result makes some people think of heating their homes entirely by solar collectors, but that ignores the cost-benefit ratio," Hanson warns. "If you size a system to supply half your yearly needs, you'll get practically a 100 percent benefit from your investment. But increasing the number of panels to supply *all* your needs may give only a 10 percent benefit for the added cost. That's because the larger system will supply more heat than you can use during most of the year."

Hanson built and sold 100 space

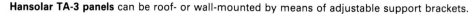

Hansolar TA-3 panels can be roof- or wall-mounted by means of adjustable support brackets.

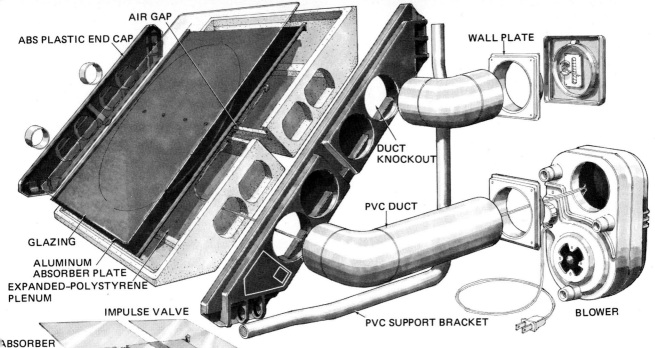

AIR GAP

ABS PLASTIC END CAP

WALL PLATE

DUCT KNOCKOUT

PVC DUCT

GLAZING

ALUMINUM ABSORBER PLATE

EXPANDED–POLYSTYRENE PLENUM

IMPULSE VALVE

PVC SUPPORT BRACKET

BLOWER

ABSORBER PLATE

CORE

VENTS (TOP & BOTTOM)

SOLAR COLLECTOR

END CAPS

PLUMBING INTERFACE

CONTROL UNIT

HEAT EXCHANGE CANNISTER

TANK

SURGE TANK

PUMP

In the space-heating version, a blower pushes cool room air through the bottom pipe into the lower plenum. Here it's blown up against the bottom surface of the aluminum absorber plate. Picking up heat as it goes, the thin film of air touching the plate rushes uphill through a narrow gap into the upper plenum. From here, the heated air moves through the upper pipe back into the house. Heating continues as long as the plate is above 70 degrees F; this keeps the blower running.

Prototype water heater uses the same parts as air heater except for the absorber plate, which is made of high-temperature plastic. With its batch-type heat exchanger, the system costs $1495. When the thermostat in the collector is warmer than the thermostat in the heat-transfer water, pump pushes a 2-gallon charge up into the collector. Water moves through channels in the absorber plate, where it's heated until a timer lets it drain back into the heat exchanger. Cycle repeats until all water is heated.

Installation of a pair of hot-air collectors takes a beginner about four hours. First, mark position of mounting brackets for panel farthest from point of air entry into house. Then slide brackets into panel and mount on wall.

Slide brackets into second panel and fit that onto connecting rings inserted into side of first panel.

Fasten brackets of second panel to wall.

Finally, connect PVC pipe or hose from second panel to house and seal all joints with silicone caulking.

heaters that first year, 1979. As his second step he listened carefully to his early customers describe their experiences installing the system and asked how they liked its modular design.

With user comments in mind, Hanson turned out 250 units the second year. This time he sought customer reactions to the system's appearance—especially the collector. Based on these reactions he put together 300 systems the third year. Since then he's built 6000 panels with only minor changes.

Hanson tackled the air collector for space heating first because, as he put it, "I'd rather save 50 percent of the 80 percent it costs to heat the average home, then save 50 percent of the 20 percent it costs to heat domestic hot water."

His design for the air collector started with the observation that only the thin film of air in direct contact with the absorber plate gets hot. The rest of the air in the collector warms up by mixing later with the hot air.

To increase the volume of air that actually touches the absorber plate and to keep colder air from diluting the hot air, Hanson devised a two-plenum collector. When the absorber plate reaches 80 degrees F, a thermostat turns on the system's blower. Cool air drawn from the house enters the lower plenum and flows in a thin stream along the underside of the absorber plate, passing through an air gap and into the upper plenum. From there it flows back into the house at the rate of 90 to 100 cubic feet per minute.

At 21 pounds the Hansolar TA-3 is a flyweight among collectors. What makes that possible is its main structure, two plenums of injection-molded expanded polystyrene. Plastic end caps, cover glazing, and support brackets keep weight down, too.

New hot-water heater

Following his usual development method, Hanson built 300 prototype water heaters this summer. Continuous production will start next spring, using feedback from owners.

A glance at Hansolar T(H$_2$O) tells you right away that something is different: Only one tube joins the collector and the heat exchanger. That's because the unique water heater uses a batch process instead of a continuous two-pipe loop.

"We studied hot-water use," Hanson explains, "and found that most homes use not more than 4 gallons in one shot. Then we looked at hot-water collectors and found that generally they operate by continuous flow during the morning and behave like batch processors in the afternoon, when most hot water is used.

"So we said, 'Why not batch all the time?'" Hanson continues. "Normally a heat exchanger holds ¾ to 1 gallon of water, which, when it gets hot, just sits there until you open the faucet. Then cold water comes in to replace the hot. So you get one gallon at 120 degrees [F] and one at, say, 60."

To solve that problem, Hanson's inner (surge) tank holds 4 gallons of domestic hot water. The outer one holds about 28 gallons of heat-transfer liquid—ordinary tap water or chlorinated well water. Two-gallon batches of liquid shuttle between canister and collectors, their stay in the collectors determined by a timer. When the canister liquid reaches 120 degrees F, the system stops, regardless of the timer. Drawing hot water lowers the canister temperature, and the shuttle resumes.

If the collector temperature falls below a setting predetermined for your area by Hansolar's computer, the system stops. Thus, at night the heat-transfer liquid drains down into the canister, where it's safe from freezing. Since the liquid has no antifreeze, the heat exchanger has a more efficient single wall instead of a double wall for safety.

In a standard T(H$_2$O) hot-water system you couple two panels to one canister. But you can improve on recovery time, which ranges from ten to thirty minutes, by adding as many as six more panels. And if you want more hot water on hand, you can connect more canisters.

Like the space heater, Hanson's solar water heater is simple, small, and light—making it inexpensive and easy to install. Filling a collector with water increases weight to only 39 pounds.

Hanson has 1000 space-heating systems on hand and the capacity to turn out more at the rate of 200 a day. As explained earlier, only 300 prototype water heaters have been built.

For a solar planning kit that includes product information, installation instructions, and an owner's manual, send $2 to Hanson Energy Products, River Road, Newcastle, ME 04553—*Erik H. Arctander*.

insulation and weatherstripping

Hardly anyone needs to be sold on the idea of weatherstripping and insulating his home anymore. Still, when I look at new or recently insulated homes, I often see a lot of areas that have been overlooked—areas where heat flows readily to the great outdoors—even though the house may have super-insulated R-30-plus ceilings and R-19 walls. Now, with the help of new materials and some old standbys, you can upgrade your insulation whether it's recently installed or ancient.

The numbers game

The R-value is an index of a material's ability to resist heat flow—the higher the number, the better the insulating ability. The R-value is used to designate how much insulation is required for specific applications in specific areas.

Manufacturers of insulation products are required to include the R-value of each item on its label. That's important information. The thickness or weight of a material is not necessarily an indication of its ability to insulate. It is the R-value rating that tells you about the material's insulating performance—and that's what counts for our purposes here. Never purchase an insulation product without first checking for the R-value on the package.

If you're looking at windows and glazing materials, you may be confronted with another number—the U-value. This is the reciprocal of the R-value, 1/R, and it's confusing because the lower the U-rating, the better the material resists heat flow.

How much R is enough? That depends on your climate. Your local weather station and perhaps your radio or television station can give you the degree-days (the difference between 65 degrees F and the mean daily temperature, added daily to give monthly and seasonal totals) for the past heating season in your area. While the accompanying chart provides broad guidelines for total attic and ceiling insulation, your local public utility or insulating contractor will make more accurate recommendations, tailored to your geographical area.

R-values are cumulative. Once you know how much you should have, measure what you already have and allow about R-2 per inch for vermiculite, or R-3 per inch for mineral wool, glass fiber, or cellulose. The difference is the R-value you need to add.

After walls and ceilings have been insulated, look for forgotten areas. One prime choice (above): where foundation wall meets exterior walls. Seal this area—the soleplate—with caulking or foam sealant, then tuck glass fiber batts along band joist. On walls where floor joists butt against band joist (below), cut batts slightly larger than openings, tuck in place. Friction fit holds them.

INSULATION GUIDELINES

Degree days	Recommended minimum R values
0–1000	19
1001–2500	22
2501–6000	30
6001 and up	38

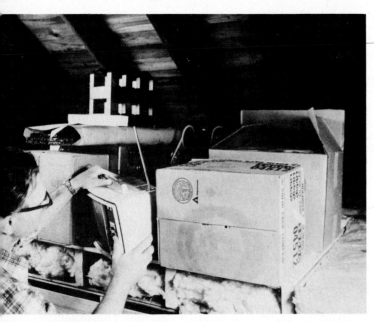

Here, joist depth did not allow for additional layer of insulation. In this case, joists could be built up and insulation laid over existing floor. Second floor could be laid on top.

You can create a storage nook by attaching plywood "flooring" to low-level collar beams, as shown. There may be advantages in placing this nook near your attic entry.

Foundation insulation

Let's say your attic and walls are insulated to the recommended maximum for your area. What next? The best advice may be to look below.

Basement and crawl-space heat loss is often overlooked on the assumption that these areas are kept warm by heat lost from the furnace or boiler and ducts and pipes. Basement heat loss costs just as much as that lost in the living areas, and reducing it results in a warmer basement and warmer floors in the rooms above.

Insulating between floor joists has been a widely accepted practice for homes with electric resistance heat. However, a University of Illinois report indicates that this may not always be a good idea; in some homes, such insulation could increase cooling costs by preventing the radiation of summer heat to the cooler ground surface.

A better idea for homes with warm-air ducts or heating pipes in the basement or crawl space: Insulate and seal the basement foundation walls and keep the heat in. You get double-duty from most of this heat; after warming the basement, it rises to warm the floors of the living areas.

Masonry foundation walls offer little resistance to heat flow, even if below grade for the first few feet. Foundation walls can be insulated from inside or out. Outside, glue or nail on a water-resistant polystyrene foam board. It should go down at least 2 feet below ground surface. Metal flashing, used at the top, diverts large amounts of water. When completed, the board should be covered to make it more attractive (stuccolike mortar looks especially nice) and fire resistant.

Inside, you can create a finished wall by building a 2 × 4 wall partition immediately inside the masonry wall—insulating it and installing a vapor barrier as you would with any conventional exterior wall—and finishing the interior in any way you wish. Better still, you can purchase one of the foam materials with drywall bonded to one surface now available. One of these, the R-Max Therma-Wall system even includes a metal mounting system that speeds the job and results in a high R-value wall complete with vapor barrier.

Sealing out winter air

While insulation will block heat from escaping from the living areas through walls and ceilings, there is another problem you have to contend with: heat that gets out constantly through cracks and tiny openings and intermittently when doors are opened—and the cold air that comes in through those same openings. According to tests by Texas Power and Light Co., up to 40 percent of the energy needs of a typical home can be attributed to air infiltration. It's sneaky, entering through barely noticeable openings—a $\frac{1}{16}$-inch crack along the sides and top of a doorway, for example, which is equal in surface area to a 4 × 4-inch hole in the wall. Seeking out such openings pays big dividends in reducing heating bills.

But what do you do when you find them? Fill any opening more than 1 inch wide with insulation, then cover it with finishing material or plywood. Foam insulation-sealants do a fine job on openings $\frac{1}{4}$ to 1 inch wide, filling the void and sealing the opening. Caulking works well on cracks up to $\frac{3}{8}$ inch wide.

Proper preparation is important for lasting results with foam or caulk-type sealants. Remove any previous caulk or paint with a putty knife and wipe the surface clean before applying the new sealer. Masking tape can be applied on both sides of the joint for a neat appearance, and it should be removed immediately after caulking, before the material begins to set up. Be sure to apply material so it contacts both sides of the opening.

Some specialty weatherstripping materials are available at hardware and building-supply stores for a variety of specific needs—replacing the seal strips around doors and windows, for example. And some of the newer materials are far better than original equipment seals on some older fixtures.

Beyond the basics

With your attic, walls, and foundation walls insulated to maximum practical effectiveness, you've undoubtedly made a big dent in your home heating bill. But there are other worlds to conquer for maximum savings, and you can start

When adding insulation to an already insulated floor, be sure to use unfaced insulation (without vapor barrier). Otherwise, batts with vapor barriers would trap unwanted moisture.

with the areas I've seen that the builder or insulating contractor forgot.

The Texas Power tests of air infiltration made a surprising discovery: The largest culprit, accounting for 25 percent of all air leakage, is the area where the frame and foundation of a house meet on an exterior wall. Known as the *soleplate*, this is the board that rests on, and is attached to the masonry foundation wall. That's a good place to start. Using caulk or foam sealant, run a bead of material around that joint for the entire perimeter of the house.

Just above the soleplate is another often overlooked area—where the floor joists attach to a perimter board that "boxes in" the ends of the joists. This is known as the band joist, and it's easy to insulate by cutting fiberglass batts slightly larger than the space between joists, then tucking them in. The friction fit holds them in place.

While you're in the basement, check every area where utility pipes and wires pass through the foundation wall—at electric meters, outdoor faucets, clothes-dryer vents, and so on. Use caulk to seal small openings, foam sealants for larger ones.

Up in the living area, consider using storm windows or doors—in severe climates, they're cost-effective even if you now have double-glazed windows and doors. Some new windows are now on the market with a special invisible reflective film between two glass panes. They have better R-values than triple-pane glass at about the same cost and weight of a double-pane. Covering doors and windows with an insulating blind or shade, or with foam-core shutters is the best insulation of all for these areas, but it requires manual operation to close them each night, and they can be cumbersome and bulky.

An uninsulated wall between an unheated garage and the house is a common source of heat leak, and it's easy to conquer. Between the exposed studs in the garage, install fiberglass batts, with the vapor barrier facing in toward the house, then cover the wall over the insulation on the garage side. I recommend an insulated garage door for any heated or basement garage. I've found that they have much better sealing characteristics, in addition to their ability to reduce heat loss.

Attic fan cover will reduce your heat losses

Using a whole-house fan instead of your central air conditioner in summer is a proved way to reduce your energy consumption. Come winter, however, and the ceiling louvers of your whole-house attic fan are likely to provide a path for the heated indoor air to escape to the attic, or for cold attic air to leak into the living areas of your home. The fan louvers are thin metal, and they often don't close as tightly as they should.

Now Manco, Inc., has made available a kit that seals off the ceiling louvers. The Manco Whole House Attic Fan Cover uses the same materials that are in the Manco Inside Storm Window Kits.

The attic fan kit contains 208 inches of self-adhesive plastic frame, spline, and enough vinyl sheet to cover a large fan. All the materials can be cut with scissors or a razor blade. The cover mounts on either the ceiling or on the fan frame. When mounted on the ceiling, the cover helps to seal off the crack between the louver frame and the ceiling.

The sheet of clear, 4-mil vinyl is secured to the frame with a plastic retainer strip that locks in the track of the frame. Excess vinyl sheet is trimmed from the outer edges, leaving about an inch on all sides (this allows the sheet to be removed in the spring and reused next winter). Except for the sheen of the plastic, the fan cover is barely noticeable against a white ceiling.

The fan louvers I covered were not very large, and there was material left over—enough to seal off the seldom-used attic hatch cover. The Manco fan cover kit sells for about $13—*Kay Keating*.

Stop air leaks around your air conditioner

Through-the-wall air conditioners are a prime source of winter heat loss if not properly weatherized. In the past, the only sure way to seal such units was to wrap the exterior portions with insulation and plastic—not an easy job, especially for units used on the upper floors and, certainly, the results were not very attractive.

Now, air-conditioner covers designed for inside mounting are available to stop air leaks. Easy to install, they attractively camouflage the unit during the winter months and are reusable year after year.

Winter Shield, about $13, is from Mortell Co., Kankakee, IL 60901. Made of polystyrene foam, it is attached to the wall with adhesive caulking strips supplied with the unit. Its depth is 4 inches and inside dimensions are 27 by 17 inches. The dark brown wood-grain finish can be painted to match wall color and, if desired, a picture can be mounted on the front.

The Hugger Comfort Cover, about $22, is from Hugger Products, Box 1072, Madison, WI 53701. It is 26½ inches wide, 18 inches high, and is available in depths of 4.3 or 9.3 inches. The off-white, high-impact plastic cover is held tightly against the wall with side clips. A foam seal tape that is pressed around the flange before mounting prevents air leaks around the edges. The installation kit, directions, and foam seal are included.

Light switches and receptacles can be a source of significant air infiltration. Rubber gaskets that go under the cover plate are available at hardware and building-supply stores. Be sure to turn off the electrical circuit before removing the cover plates to install them, however. If you're choosing new light fixtures, don't consider recessed or unvented fixtures for an insulated ceiling. The National Electric Code requires that insulation be kept a minimum of 3 inches away from such fixtures, leaving a sizable void, where heat could escape rapidly. Hanging fixtures and track lighting avoid the problem. (Insulation must also be kept away from flues.)

Before insulating, it's a good idea to have a qualified person or electrician determine that your home's wiring system is in good condition. If you have experienced problems—excessive dimming or flickering lights, frequent tripping of fuses or circuit breakers, sparks or glowing from outlets, or overheated outlets or cover plates, for example—have the system checked and repaired immediately.

Foam sealants prevent air infiltration in large holes, such as for air-conditioner hoses. When applied outdoors, foam sealants should be painted with latex enamel to prevent deterioration.

Insulation contractors can blow-in insulation. They wear face masks to prevent respiration of airborne fibers. Contractors should offer an estimate and give evidence of insurance coverage and professional recognition.

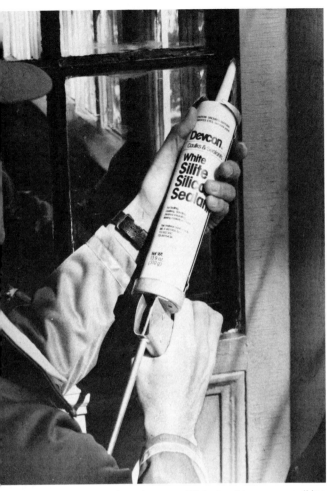

This window has Heat Mirror film and is as transparent as any double-glazed window. It has **48** percent higher R-value than a triple-glazed unit. Material's atom-thin metallic layer reflects heat like foil, while retaining transparency.

Above, for holes and cracks up to 3/8-inch wide, use a caulking gun to apply silicone sealant. Below, you can install a do-it-yourself storm window by Stanley. Here self-stick edging is applied; then film is positioned and locked in place with a sealing strip.

A vapor barrier is important to prevent moisture generated inside the home from penetrating the walls and entering the insulation. This is usually formed by backing on the "warm-in-winter" side of insulation, or by a plastic film applied over stud walls beneath the interior wall material. If your home doesn't have a vapor barrier, you can paint one on—literally—with a new type of paint that forms an effective protective coating. This paint should be used only on top-floor ceilings and exterior walls; conventional paint is applied right over it.

In the attic, check the condition of eave and gable vents to be certain that newly installed insulation is not blocking them. Proper ventilation over the insulation is important in order to remove moisture in winter and heat buildup in summer. Never cover roof vents in winter on the assumption that you're keeping heat inside—you may only be saturating your insulation.

Here are some precautions to keep in mind when you're working with insulation: (1) Wear a long-sleeved shirt and long pants, a dust mask, safety goggles, a hard hat, and gloves. (2) Don't smoke. (3) If you're working in an attic, use a walk board placed across joists, since the material between them won't support your weight. (4) Provide good lighting. (5) Beware of nails.

A thorough insulation and sealing job does much more for your home than save money—it improves the comfort factor many times over. Warmer glass surfaces eliminate the "draft" often felt near cold glass and allows you to keep the humidity higher without condensation. Good sealing combined with good insulation eliminates drafts and reduces noise levels. Best of all, these benefits are reaped year-around as your home keeps out summer heat and moisture just as effectively as it retains them in winter—
Evan Powell.

traditional american cupboard

Some furniture pieces, more than others, speak truly to that which is home. A cupboard of plain board lumber worn to satin with use is one of them.

Originally a small, utilitarian closet with a cabinet for food storage and open shelves for dishes, cupboards saw timeless service in Early American kitchens. Later, cupboards got fancier and moved into the dining room or hallway. Now, regardless of any particular stamp a cupboard is a distinguished piece—always popular and still useful for display and storage.

Our Traditional American Cupboard is a personal design that follows general cupboard lines but also includes some nice richness of detail through raised panels, moldings, a simple curved embellishment near the top, and small articles of hardware. The cupboard also is designed for basic construction in the home workshop with standard power tools. You can have this edition in two or three weekends of sizing, cutting, assembling, and finishing.

We went for mahogany throughout. The construction plan itself actually is two pieces—the base cabinet and the top network of open shelves.

We began by first gluing up ¾ × 5½-inch mahogany stock to form the cupboard's top, sides, and two shelves. Edge the boards square on a jointer-planer and then cut ⅜ × ⅜-inch rabbets along the sides of each board for a better glue joint. The boards are then edge-glued and bar-clamped together to dry.

Across each end and across the middle, we C-clamped 2 × 3-inch boards to keep the glued-up stock flat while under the pressure of the bar clamps. After the glue has dried for twenty-four hours, the glue beads between each joint must be removed and any raised lips between the joints planed smooth.

We then used a belt sander to quickly smooth the large glued-up

Inside edges of door frames are routed to accept raised-panel inserts. The rounded corners are chiseled out square by hand.

panels. When making furniture, it is easier to sand the individual pieces than to try to sand the complete piece after it has been put together.

The door-and-drawer frame comes next. We first rip-cut the ¾-inch stock to size and then put the frame together with ⅜ × 2-inch wood dowels and glue. Before applying glue, however, put the frame together dry and bar clamp in place to be sure all parts fit together correctly. This may seem like extra work, but one incorrect board throws the entire frame out of square.

The center door divider is ½-inch stock glued in place. Across the top and bottom of this divider

Raised panels are cut on radial-arm saw with the blade set in the bevel-rip position. Bevel angle should be 74 degrees.

¼ × 2¼ × 5½-inch strips of mahogany are then glued in place—with the grain running horizontally. The center raised panel, which is cut later, is then glued and C-clamped onto the ½-inch center board.

The door frames that hold the two raised panels in each door are made of 2½-inch wide stock put together with 2-inch dowels and glue and bar-clamped to dry. When making doors, it is essential that all boards be free of warpage. Warp causes any door to fit incorrectly when hinged.

When the glued frame has dried, ¼-

Center door-divider goes into door-and-drawer frame (backside shown) and is secured with strips glued top, bottom.

inch deep-by-⅜-inch wide rabbets are cut with a rabbeting bit-and-guide around the inside front edges—these to accept each raised-panel insert. After routing, the rounded corners must then be chiseled square before the raised panels fit.

We cut the five raised panels next. Each panel is cut from ¾ × 5½ × 17½-inch stock, using a radial arm saw. We first constructed a jig to secure the panel while pushing it through the blade. A new guide fence also must be made to accommodate the blade in the bevel-rip position; the back fence pieces must be removed for the tilted blade.

To cut the panels, secure the boards

Reprinted from *Mechanix Illustrated* magazine. Copyright by CBS Publications.

in the jig and set the saw in the bevel-rip position at a 74-degree angle. Then lower the saw until the outside edge of the beveled panel measures ¼-inch and the panel bevel is 1½-inches wide.

By using a carbide-tip blade, the bevels can be cut in one pass. The ⅛-inch thickness of the blade tips form the upper edges of the panel at a 74-degree angle—which you then sand to a 90-degree angle. This eliminates the need for making four additional cuts at 90 degrees.

Before cutting, be sure the blade moves freely, as the cut of the blade is lower than the table surface. Make the first cut with the grain, then remove the jig arm and make the next cut across the grain. Replace the arm and make the next cut across the grain.

Reposition the arm and make the third cut with the grain; then make the last cut across the grain. The panel is then sanded to remove any saw marks. Each of the five panels is now glued in place.

While you have the saw set up for cutting raised panels, cut the panel for the drawer front. At the same time, you also can cut the beveled edge that forms the shelf top. When cutting this 74-degree angle, raise the saw blade ⅛ inch so that you have a flat bevel without a ⅛-inch lip.

Next, cut ⅜ × ¾-inch dadoes across the insides of the cupboard's two sides for the two shelves. The bottom dado also is cut across the back of the door frame for the bottom shelf. Do not cut dadoes across the front frame for the center shelf.

Cut ⅜ × ⅜-inch rabbets along the inside edges of the front frame and the two sides. A ¼ × ⅜-inch rabbet also is cut along the inside back edge of the two sides for the ¼-inch plywood back. We used a T-guide clamped to the sides to cut the dadoes and a router guide to cut the dado across the bottom back edge of the door frame.

Top piece with shelves installed (minus back panel) is squared. Then, strips are nailed on the shelves as plate retainers.

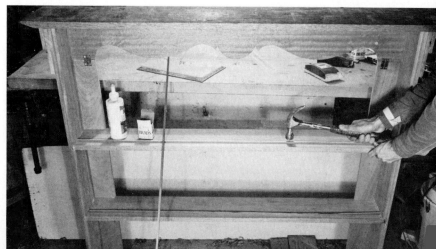

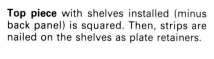

Glue along the dadoes and rabbets and connect the two shelves and the front frame to the sides. The five pieces must then be bar-clamped together to dry. Metal corner braces can be screwed to the undersides of the two shelves for added support.

While the cabinet dried, we fashioned the drawer guides. They are made of ¾-inch stock and put together with glue and wood screws.

Each guide is then secured to the cabinet sides and front with glue and 1¼-inch wood screws. The top drawer guide is secured to the cabinet top with glue and screws after the top has been connected to the cabinet base with 1-inch metal corner braces and screws.

The drawer for the cabinet is made of ½-inch mahogany and ¼-inch lauan mahogany plywood for the bottom. The raised panel drawer front is glued and nailed to the ½ × 4⅜ × 38¾-inch board—allowing ½-inch on each end and ¼-inch across the bottom edge for drawer sides and bottom. The drawer sides, bottom, and center divider are glued and nailed in place after squaring the pieces.

The cupboard display shelves are made of ¾-inch and ½-inch mahogany. We first cut the two scrolled top pieces with a saber saw. The sides and shelves were cut next—with ⅜ × ¾-inch dadoes in the sides for insertion of the shelves.

To make the shelves' front edges fit flush with the front side rails, ½ × 3⅝-inch notches were cut in the three shelves. Before gluing the shelves in place, cut a ⅜ × ½-inch rabbet along the inside back edge of the two sides for the ½-inch stock that forms the back. Glue and bar-clamp the shelves in place.

The ½ × 4-inch front side rails are glued and nailed in place, as is the 5½-inch scrolled section. The smaller scrolled piece is then glued and C-clamped on top of it, as are the side pieces. The top, made earlier, is now nailed in place.

The shelf back is made of ½ × 9¼-inch boards that are nailed in place. To keep plates from falling over, ¼ × ¼-inch strips are glued and nailed 1-inch from the back of each shelf.

To finish the shelves, we cut a ¼-inch bead-and-cove edge, with the router and guide, along the underside of each shelf. Countersink all nail holes and cover with putty.

We gave the cupboard a finish coat of Watco Dark Walnut Danish Oil and went over this with two coats of tung oil. When the finish dried, the hinges and handles were attached—*Bruce Murphy. Photos by Bruce Blank.*

Back of top piece is not sheet material. Instead, ½-inch mahogany boards, each 9¼ inches wide, are used for their butt-grain effect.

queen anne table

Modeled after the fine pieces made in eighteenth-century England, this Queen Anne tea table is assembled with a simplified, modern technique using ready-made cabriole ("Queen Anne") legs. For the craftsman who would rather make his own, we show a pattern in a squared drawing. The pattern can be modified, of course, by the more experienced craftsman who enjoys creating an individual design.

No matter how you change the design of the leg, remember the axiom that a vertical centerline through the top of the leg should intersect the center of the foot.

Cut the four aprons to size and shape, miter the ends, then glue and clamp them together on a flat surface. Be sure to let the glue set completely, as it will be the only thing holding the corners together until they are reinforced with the legs.

Cut the legtops as indicated, unless you use the optional legs that attach with plates, and glue and screw them inside each corner of the apron assembly as shown. Reinforce them with the glue blocks as indicated.

Make the two frames, one of picture-frame molding (purchased or made as detailed) and one of ½-inch stock. Glass is fitted in the rabbet of the molding—this should be plate glass at least ¼-inch thick—then the molding

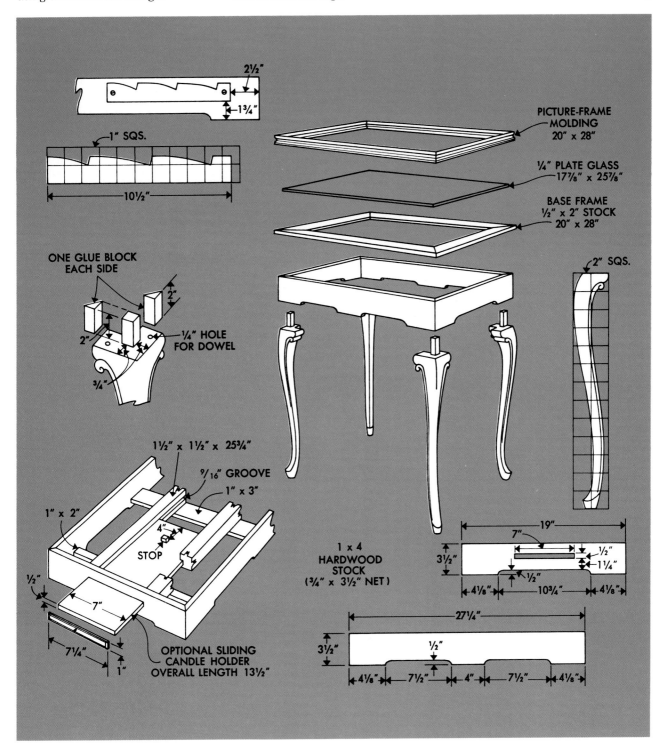

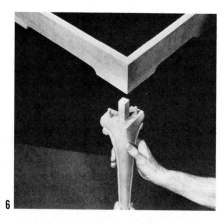

5

Matching holes are drilled in leg tops, with legs held in vise. Legs are padded to prevent vise damaging them.

Top can be tilted up to display reference materials.

6

Place dowels in legs, check fit to aprons. Make any necessary adjustments to assure perfect fit between legs and aprons.

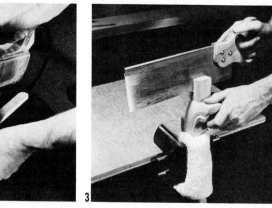

1

3

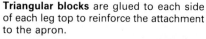

7

Profiles on lower edges of aprons are cut with portable jig saw, pieces saved. Jig saw or band saw can be used instead.

If ready-made legs are used, tops are cut to provide "shelf" ¾ inch wide on which adjacent aprons rest.

Projections on leg tops are cross-drilled for No. 8 × 1-inch brass screws. Glue is applied, screws driven into aprons.

Ends of aprons are mitered, glue applied, then assembled and held with strap clamp. Check for square corners.

Holes for ¼- or ⁵⁄₁₆-inch dowels are drilled in lower edges of aprons. Thumb tack makes dowel-center tool to mark holes.

Triangular blocks are glued to each side of each leg top to reinforce the attachment to the apron.

2

4

8

9

Pressed-wood ornament is located at center of each long apron ¼ inch above lower edge. Use glue and small brads.

13

Optional leg attached with metal plate is attractive and functional; not quite as graceful as leg used on original.

10

Picture-frame molding (made or purchased) is attached to frame of ½-inch stock with screws driven into molding.

14

Three of the screws that hold metal leg plate are driven into lower edges of adjacent aprons. Predrill holes.

11

If you wish top to tilt up for display or for convenient reading or study, attach brace with ¾-inch brass butt hinge.

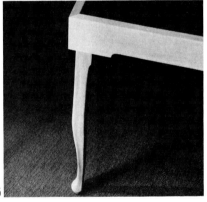

15

Optional leg is ready-made type that fastens by means of metal plate. In this case, glue and screw triangular pieces.

Notched hardwood bracket is screwed to inside of one or both side aprons to accept end of brace that holds up top.

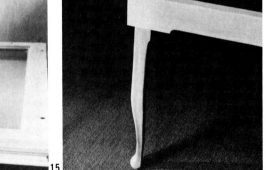

12

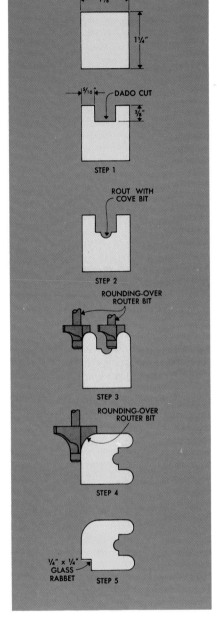

STEP 1

1⅛″   1¼″

⁵⁄₁₆″ DADO CUT ³⁄₈″

ROUT WITH COVE BIT

STEP 2

ROUNDING-OVER ROUTER BIT

STEP 3

ROUNDING-OVER ROUTER BIT

STEP 4

¼″ x ¼″ GLASS RABBET   STEP 5

is screwed to the frame of ½-inch stock.

If the top is to be tilted so you can use it for display or easy reading, install two hinges inside one long apron and on the long side of the frame assembly as indicated. Also install the prop and notched bracket. Position the prop to hold the top at the three angles you wish.

If the optional candle trays are installed, replace the glass with a piece of ¼-inch hardwood paneling so the "mechanism" is hidden.

Stain and finish to suit your decor. Apply a number of coats of varnish or lacquer for a deep, rich finish—*John W. Sill.*

three credenza styles: one basic plan

Furniture-making, like any other applied skill, challenges you at each new level of advancement—usually, the bigger and seemingly more complicated projects. Credenzas, for example, are important as well as massive pieces that address themselves to this point. They scare a lot of woodworkers who then miss the excitement of expanding on their skills.

Why is not as important as how; a good, slow studious look at what's involved usually brings a sigh of relief with an understanding of the techniques. One step is sure to follow the other.

To show that advanced, as well as diverse, styles are yours for the study, we have constructed a trio of credenzas virtually from one set of plans. Each is faced on a common frame, even though all three are representative of different historical periods.

Our basic credenza is Early American, in cherry, with raised-panel doors and thumbnail molding around the top. For ease of building in small-home shops, the frame is made in several easily-handled assemblies: base, floor, partitions and ends, top frame and top.

Using the same basic frame the traditional oak credenza was made with a few changes. To save money the floor was replaced with a frame identical to the top frame. Floors glued from scrap were used in the wings. Three flush drawers replaced the doors and hidden drawer in the center section.

The third credenza is English Regency, a style that came along after Queen Anne, Chippendale, Hepplewhite, and Sheraton. It was the style of the early nineteenth century, adapted by the great American furniture-makers, such as Duncan Phyfe.

This regency mahogany credenza is actually closer to the basic cherry credenza in construction than is the oak one. The only change other than the doors is the addition of a 1½-inch-wide apron across the top under the frame molding, plus different molding on the frame and floor, and none on the top.

All three credenzas are constructed of hardwood lumber, which can be mail-ordered from Educational Lumber Co., Inc., Box 5373, Asheville, NC 28803. About 50 board feet of ¹³⁄₁₆-inch oak and mahogany, or ¾-inch cherry lumber is required for each credenza. An exact amount depends on how the random width and length lumber runs and how much waste you end up with. Half-inch oak is used for the drawer sides and backs.

The construction of the Early American cherry credenza is described

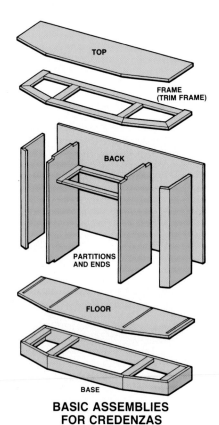

TOP

FRAME (TRIM FRAME)

BACK

PARTITIONS AND ENDS

FLOOR

BASE

BASIC ASSEMBLIES FOR CREDENZAS

first, with notes following on the changes required for the other two credenzas.

Early American cherry

Base. The base has six pieces—center, wings, sides, and back—plus internal framing. Rabbet and dado the back into the sides. Assemble the odd-angled front joints in either of two ways. Reinforce the mitered end joints with glue and blocks, or with splines. Dowel or spline the center joints with a reinforcing block glued or screwed on the inside. Splining is preferable because the spline locks the joint in alignment. Birch aircraft plywood in ⅛-inch or ³⁄₃₂-inch thickness makes excellent splines, and it is available in small pieces at better hobby shops if the yard doesn't stock it.

Begin by blanking the six base pieces to width and slightly over length. Set your saw for a 7-degree angle and trim both ends of the center piece and one end of each of the wing pieces. Without changing the angle, lower the blade and cut kerfs for the splines using a tenoning jig. Reset saw to a 38-degree angle and trim the other ends of the wings and one end of each end piece; cut kerfs for splines. Cut splines; clamp the five pieces together and check your joints. If angles are slightly off, adjust the length of the back piece slightly to get a good fit on the front angles rather than to try to correct the angles. When the back length is determined, cut it to length and rabbet, then dado the end pieces to receive the back. Make the center glue blocks and the parts for the T-shaped cross frames. These frames are doweled to the back; drill the dowel holes in the back.

From ¾-inch plywood or scrap make an accurate 14-76-90-degree triangle about 12 inches long for jig to help get the base and frame joints lined up accurately. Start base assembly by gluing the ends and wings together. Next, glue and nail, or dowel the glue

EARLY AMERICAN CHERRY CREDENZA demonstrates the basic credenza construction. To form the curved bevel on the door panel, rout outline of raised panel area, remove waste with gouge, then sand smooth and to dimension with a sanding disc in a hand drill.

TRADITIONAL OAK CREDENZA replaces the center doors and hidden drawer with three flush drawers. Here the oak panel is set into assembled frame. Shape molding with a combination of router cuts, then sand smooth, cut to length, and nail in place.

REGENCY MAHOGANY CREDENZA duplicates the cherry credenza, but adds decorative period details with moldings and four doors. Use router to make the half-round molding from ¼-inch mahogany stock and glue in place.

blocks to the vertical cross frames. Glue and clamp the back between the ends and glue the center piece to the wings. Clamp all the parts to your bench. Immediately glue and screw glue blocks to center and wings. Line up the vertical frames against the back and drill through for dowels. Glue and insert dowels. Check alignment before the glue sets.

Fit the top horizontal pieces of the center frames, then glue and screw them to the vertical frame pieces; add glue blocks, and dowel them to the back. While drying, cut and fit end cleats; glue and screw cleats to the ends. Base complete, except for sanding.

Frame. The frame, like the base, has six pieces—center, wings, ends, and back—plus two internal cross pieces. Front joints can be doweled (watch the angles so you don't end up with something impossible to assemble), mortised-and-tenoned, or splined. To spline, the mortises have to be stopped on the outside, making it necessary to rout or handcut them. I used splines, and mortised with my router set up as a shaper.

Blank the six frame pieces to width and slightly overlength. Miter the ends of the center, wings, and end pieces and mortise for splines. Since the back is tenoned to the ends, mortise ends for these tenons also. Next, mortise the back and center pieces for the cross frame tenons. Note that these cross frames must be positioned accurately as their outer edges position and provide attachment for the partitions.

Frame assembly is not easy because it's awkward getting clamping pressure applied to the odd-angled joints. Exactly how you do it depends on your clamp supply so proceed cautiously. Begin gluing the center, cross frames, and back together, with the rest of the frame clamped-up dry as a jig to assure correct alignment of the glued joints. When dry, glue end and wing on—one side at a time. Note that the

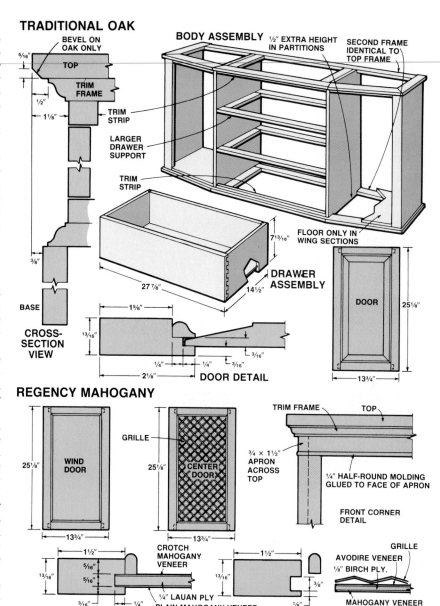

Block frames up and off your workbench, then rout molding on the two oak frames simultaneously.

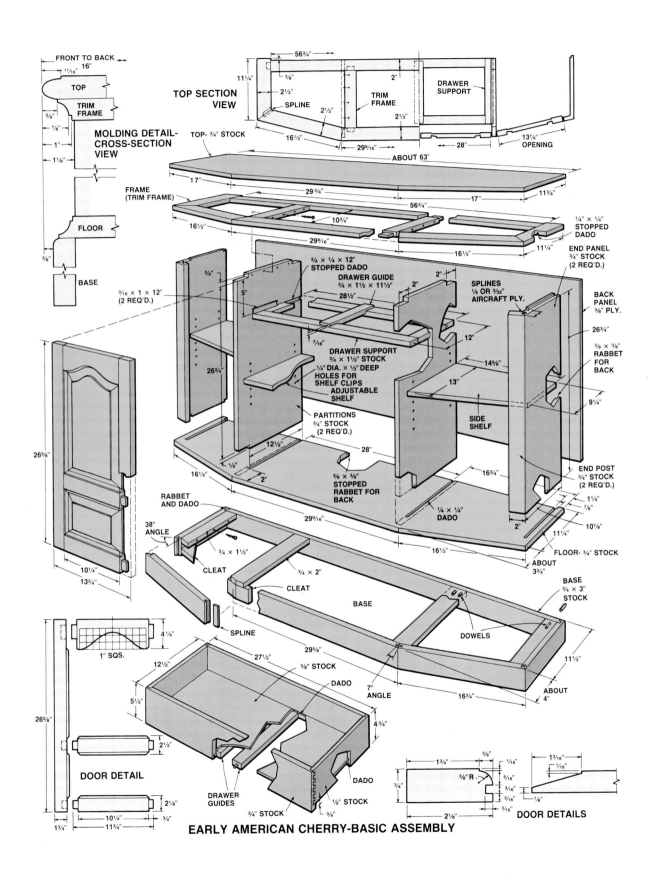

FRONT TO BACK
16"
11/16"
TOP
TRIM FRAME
5/8"
7/8"
1"
1 1/8"

MOLDING DETAIL-
CROSS-SECTION
VIEW

FLOOR
3/8"

BASE

TOP SECTION
VIEW

56 3/4"
11 1/4"
3/8"
2 1/2"
2"
DRAWER
SUPPORT
SPLINE
2 1/2"
TRIM
FRAME
2 1/2"
16 1/2"
29 9/16"
28"
13 7/8"
OPENING

TOP- 3/4" STOCK

FRAME
(TRIM FRAME)

ABOUT 63"

17"
29 3/4"
17"
11 3/4"

56 3/4"

16 1/2"
10 3/4"

29 9/16"
16 1/2"
11 1/4"

1/4" × 1/4"
STOPPED
DADO

END PANEL
3/4" STOCK
(2 REQ'D.)

3/4"

3/4 × 1/4 × 12"
STOPPED DADO

DRAWER GUIDE
3/4 × 1 1/2 × 11 1/2"

2"

SPLINES
1/8 OR 3/32"
AIRCRAFT PLY.

BACK
PANEL
3/8" PLY.

3/16 × 1 × 12"
(2 REQ'D.)

5"
28 1/2"

2"

7/16"
DRAWER SUPPORT
3/4 × 1 1/2" STOCK
1/4" DIA. × 1/2" DEEP
HOLES FOR
SHELF CLIPS
ADJUSTABLE
SHELF

26 3/4"

26 3/4"

12"

14 5/8"

13"

SIDE
SHELF

26 3/4"

3/8 × 3/8"
RABBET
FOR
BACK

9 1/4"

PARTITIONS
3/4" STOCK
(2 REQ'D.)

12 1/2"
1/4"

28"

3/8 × 3/8"
STOPPED
RABBET FOR
BACK

16 3/4"

END POST
3/4" STOCK
(2 REQ'D.)

1 1/4"
7/8"

10 7/8"

26 5/8"

16 1/2"
2"

29 9/16"

1/4 × 1/4"
DADO

16 1/2"

2"

11 1/4"

FLOOR- 3/4" STOCK
ABOUT
3 3/4"

RABBET
AND DADO

38°
ANGLE

3/4 × 1 1/2"

CLEAT

CLEAT

3/4 × 2"

BASE

DOWELS

BASE
3/4 × 3"
STOCK

11 1/2"

10 1/4"
13 3/4"

4 1/4"
1" SQS.

SPLINE

12 1/2"

27 1/2"

3/8" STOCK

DADO

7°
ANGLE

16 3/4"

ABOUT
4"

5 1/2"

29 5/8"

4 3/4"

26 5/8"

2 1/2"

DOOR DETAIL

DRAWER
GUIDES

3/4" STOCK

1/2" STOCK

3/4"

DADO

1 3/4"
3/8"
1/16"

1 3/16"
1/16"

3/4"

3/8" R

5/16"

3/16"

3/16"

1/8"

2 1/8"

2 1/8"

10 1/4"
3/4"
11 3/4"
1 3/4"

DOOR DETAILS

EARLY AMERICAN CHERRY-BASIC ASSEMBLY

back piece is set in ⅜ inch from the back of the ends. This provides a nailing surface for the plywood back. Check alignment again to be sure your clamps aren't pulling the frame parts out of position or bending the back piece.

Sand all pieces smooth and flat. Glue up blanks for the top, floor, partitions, ends, shelves, and drawer front.

Assemble the drawer frame. Put together the frame with either dowels or mortise-and-tenon joints. Sand the top and bottom of the frame flat and lay it on the base. Check the alignment of the wing angles. The frame is ¼ inch smaller than the base on front and sides. Now lay the frame on the floor and trace the outline; cut out floor. Trim the floor to dimension and rout the stopped rabbet in the back of the floor to receive the cabinet back panel. (Remember that the back of the frame is already set in ⅜ inch for the same purpose.)

Ends and partitions. Rabbet the top and bottom of the ends, and the bottom of the partitions to form tenons. Cut blanks for the end posts, and bevel the mating edges of the ends and end posts to form a mitered joint. Reinforce this joint with a spline. Rabbet the back edge of the end panel to receive the cabinet back panel. Route mortises for hinges in the end posts and partitions.

The molded edges of the frame and floor are the same, but inverted. Clamp both to the workbench so they can be routed at the same time—walking round and round. Doing them together saves trying to reset the router again for both for successive cuts.

Place the frame on top of the floor again, making sure top sides are both up, and trace the location of the outer

Cobble fixture to rout waste stock from cherry door frames. Two rails support router faceplate with nailed-on crosspiece for router limit-stop. Insert and clamp stile under crosspiece, butting against stop lock clamped on bench. Clean out waste with straight or rabbeting bit in router. Clamp a second stop limit across rails to clean out center of stiles.

frame edge cross frames on the floor to locate the dadoes for the partition tenons. Flip the frame over (bottom up) and back to back with the floor; align and clamp. Mark both to locate the dadoes for the end assembly tenons; rout all six dadoes.

Clamp the two partitions to your bench, aligned back to back with center sides up. Rout stopped dadoes to receive the drawer frame. Unclamp, then notch the ends of the drawer frame front to fit the dadoes. Drill and countersink holes in partitions for screws to go into the drawer frame and into the top frame cross frames.

Body assembly. Clamp one partition to frame and fasten with screws; repeat for second partition. Fasten the drawer frame between partitions with screws. Fit the bottom tenons into the dadoes in the floor. Clamp sufficiently so assembly won't fall apart, then fit end assemblies to the frame and floor.

Check trial-assembly to be sure the partition and end frame front and back edges are aligned in relation to the edges of the floor and frame at both front and back.

(Note: You need eight pipe or bar clamps for the final body assembly step. You can use six if you have heavy clamping cauls available to clamp the ends with one clamp each.) Disassemble body. Begin permanent assembly by gluing and screwing one partition to the frame. Carefully block parts so the first joint is absolutely square. When dry, glue and screw the drawer frame end to the same partition, supporting the other end, then add second partition to assembly. Glue ends to the frame, one at a time, while using the floor as a jig for alignment. (Note: Body is assembled upside down.) Last, glue floor to partitions and ends. Fill all holes in the partitions.

Place the top blank on the bench upside down and position the body on it. The top should be ⅝ inch larger than the frame on front and sides. Draw finish outline of top and cut to size; rout thumbnail molding on top edge.

Drawer. The drawer front extends to cover the drawer frame. Dovetail the drawer sides to the front; dado-rabbet the back to the sides. Fit the bottom into grooves in the front and sides and slide under the back. Dowel two center guide rails into the front of the drawer; fasten through bottom to the drawer back with screws. Check that rails are square with the drawer front.

Doors. Make the frame-and-raised-panel doors with two panels separated by a rail because a single panel is not proportioned attractively. Mortise and tenon frames together, and fit panels into grooves in the stiles and rails. Make the stiles and rails before tackling the panels.

Blank all stiles and rails in back-to-back pairs, except the center rails which are worked individually. Shoulder the rail tenons (three sides), then check tenons on the sides only. Saw and sand curve on top rails. Saw and rout grooves for panels. Use a router set up as a shaper for the groove in the curved edge at the top rail. Using a ⅜-inch rounding-over bit, rout molding on rails and stiles. Rip pairs apart.

Set saw blade at a 45-degree angle to miter the molding ends. Reset blade to a 0-degree angle, back it off a bit, and slot molding where it has to be removed from the stiles every half inch or so. Chisel out waste wood. Cobble a fixture such as the one shown

Clamp the two drawer frames between the partitions on the base for the oak credenza. Using your clamps judiciously, complete the body assembly.

in photo detail, and clean out all stile mating surfaces with a straight or rabbeting router bit. Complete shaping tenons on rails; drill and chisel out mortises on stiles. Test-assemble all doors to check fit.

With assembled door frame, trace outline of two openings on paper; adding 3/16 inch all around for panel patterns.

The straight panel sides are easily raised on a table saw. Forming the curved panel requires several steps. First, referring to pattern, make a plywood or hardboard template to guide router with guide-bushing to make initial cut outlining the top edge of the raised panel area. Next, saw the panel to rough but oversize dimension. With gouge and mallet rough the bevel. (Note: If you don't have a gouge, rough bevel with coarse-sanding disc in an electric drill.) With fine-sanding disc in drill, smooth the bevel to dimension. Test-fit the top door rail to the panel. Continue sanding until the panel fits into the groove with only 1 inch of bevel exposed. Complete the raised panel by sawing bevels on panel bottom and sides in the usual manner.

Before gluing up the door frames and panels, make sure the panels fit, but not too tightly. They shouldn't rattle, but when rapped with a knuckle, they should sound loose. Stain the panels.

After sanding, fit the doors and mortise for hinges. The open edges of two side doors must be beveled 14 degrees to lie flat against the partition. Install bullet catches top and bottom.

Traditional oak

The center doors and single inside drawer are replaced with three flush drawers. Instead of a full floor, a second frame identical to the top frame is used. Because of this, the partitions do not have rabbeted tenons on the bottom and are blanked with an extra 1/2 inch in height. Floors are in the wing sections only, placed on top of the new bottom frame. Trim strips of matching height are at bottom and top of center section and at top of wing sections. This added trim gives a more massive appearance to the moldings, in keeping with the use of oak.

The doors are constructed differently. The raised panels are held in place with separate moldings. Note the step in bottom of the frame rabbets as shown. Set panels into the frames only 1/4 inch, and the 1/2-inch-wide molding laps the panel only 1/4 inch. The shoulder also helps keep the moldings seated square. Shape moldings with a combination of router cuts, sand smooth, rip from supporting boards, and nail in place.

The drawer construction is the same as for the cherry except that there is

no overlapping lip on the bottom edge and the dimensions are changed to 27 7/8 inches width, 7 13/16 inches height, 14 1/2 inches depth. If desired, a dust panel of 1/4-inch plywood can be nailed in the bottom of the floorless center section.

Regency mahogany

The only change in body construction is the addition of a 1 1/2-inch deep apron across the front and below the top frame. The end posts are cut 1 1/2 inches short to accommodate the apron. Glue half-round molding made from 1/4-inch stock around the case. Round board edge with a router, sand smooth, then rip molding from the board.

The frames for side and center doors have different width dadoes to hold the panels, and the bottoms of the center doors can be removed, since they are attached to the stiles by screws only. Trim door frames with the same half-round molding used to trim the body, fitting inside the edge of the frame.

The side door panels are 1/4-inch lauan plywood veneered on the front with matching mahogany crotch veneer, and on the back with plain mahogany veneer. The center door panels are 1/8-inch plywood, veneered on the front with avodire, and on the back with mahogany. Thinner panels are used to allow space for the metal grille.

This bright brass grille (it also comes antiqued) is actually square steel wire brass-plated and lacquered, and it is hard work to cut. Use a jab hacksaw to cut pieces from stock, then heavy dikes to trim ends. In the sawing, notch the wire, then break it by carefully bending the whole grille. It is important when cutting the grille to have it fit symmetrically in the door, and to have both grilles identical.

To finish. Sand thoroughly. Sand everything you can before assembly—partitions, ends, door panels, interior surfaces. Make backs from 1/4-inch lauan plywood, but do not fasten to credenza.

Fill open grain of mahogany and oak credenzas with a paste filler.

Use Zar's Natural Teak No. 120 satin stain on the cherry credenza; Constantine's Brown Mahogany NGR (nongrain raising) stain on the mahogany credenza; and Sherwin-Williams's Fruitwood on the oak credenza. Allow to dry.

For final finish, apply three coats of Constantine's Wood-Glo, lightly sanding between coats. Fasten back piece in place—*Thomas H. Jones.*
Photos by Bruce Blank

COMPARISON PARTS LIST FOR THREE CREDENZAS

Early American cherry credenza	Traditional oak credenza	English Regency mahogany credenza
Top	Same	Same
Frame	Same	Same
Ends	Same	Same
End posts	Same	1 1/2 inches shorter at top
Partitions	Same (change dado for larger drawer frame)	Same
Drawer frame (1)	(2)	(1)
Floor	Replace with bottom frame, end floors, and trim	Same
Base	Same	Same
Doors (4)	(2)	(4)
Drawers (1)	(3)	Same (1)
Hinges: Amerock No. 2355	Same	Same
Door pulls: Amerock Pendant Pull No. 886	*Smith Supply, Inc. Drop Pendants No. 33B-AE	*Smith Supply, Inc. Ring Pull (2) No. 4619X3 Knob (2) No. 1588
Drawer pulls: Amerock Pendant Pull No. 886	*Smith Supply, Inc. Bail Pulls No. 71B-AE	Not used

*Smith Supply, Inc., 120 W. Lancaster Avenue, Ardmore, PA 19003

living/dining divider with storage

One of the greatest needs we face in "staying put" is increased storage space. The longer we occupy a home, the more we accumulate and the less space we have to store it in.

I needed a storage space that would also serve as a visual separation between my open-plan living room and dining room. Since the dining area is an interior space with no windows of its own, however, the divider could not obstruct light and ventilation from the living room. The cabinet shown here answers these needs neatly.

Construction begins with the interior frame (drawing); I used construction-grade lumber since it's all hidden behind the ¼-inch and ½-inch plywood that lines the compartments. The butcher-block top came cut to size; I attached it with screws, *predrilled* through the 2 × 2s from the underside. The divider is faced on the dining side with gypsum board; the living-room face is ¾-inch birch-veneer plywood for a warm, low-maintenance surface. Magnetic catches secure the doors. The clever hinges came from The Woodworkers' Store (21801 Industrial Boulevard, Rogers, MN 55374). Price: $4.50 a pair—*Richard Stepler. Photos by John Keating.*

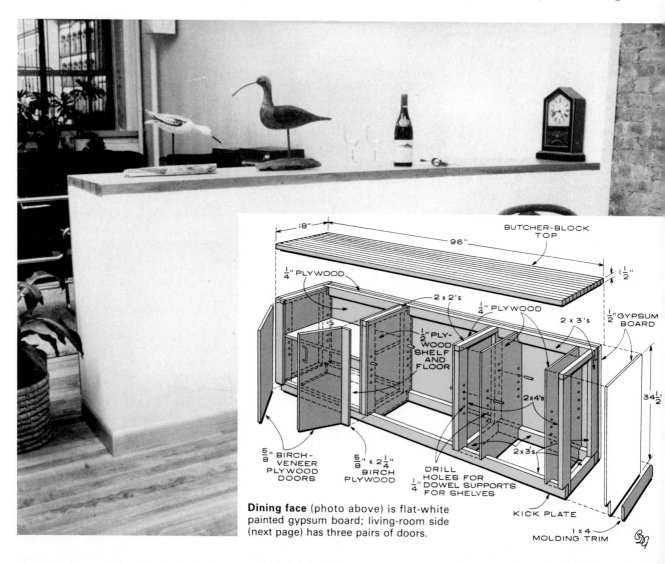

Dining face (photo above) is flat-white painted gypsum board; living-room side (next page) has three pairs of doors.

European no-mortise concealed hinges are circular, need only two 14-mm holes for installation.

child's desk

Everyone needs a special place to get away from it all—children, too. With just a little shopwork and imagination, you can create a child's hideaway in the most cramped of homes. The cabinet, desk, and bulletin board I built for my daughter Carrie are designed around the theme of trees and flowers. They give her a quiet, cheerful place to play, study, and store her treasures.

I began with a basic plywood box for the cabinet (most dimensions are unspecified so you can adapt the design). The top, bottom, and sides meet in tongue-and-dado joints; the thinner back fits into grooves in each piece. The box is raised a few inches off the floor on a black-painted square base.

The front face and door of the cabinet were made using some richly textured butternut wood that I had available. First I made and permanently attached a mitered frame to match the outside front dimension of the box. The door itself is a multilayered assembly: Its frame of butternut strips is mortise-and-tenoned to stand up to the strain it will get in use. The inside back of the frame is rabbeted (before assembly) to receive the door panel. I made the panel from tongue-and-groove stock glued up at a 45-degree angle and cut off square. The front edges of the panel were rabbeted back ³⁄₁₆ inch more than the frame rabbet to allow for expansion. For the same reason, I didn't glue the panel in but secured it with battens. Assembled, the slightly recessed panel makes an ideal setting for an applied design.

Next came the work surface: particleboard cut to fit the corner, covered with vibrant-green Wilson Art laminate, and finished with a molding strip on the visible edge. It is supported at a height of 28 inches by wood strips screwed to the cabinet top and walls. An inch of space between the cabinet and work surface allows for storage of artwork. For more storage I hung particleboard shelves.

No child's room is complete without a bulletin board. I covered the entire wall over the desk with ¼-inch cork mounted with countersunk nails and trimmed around the edges with butternut strips. The flower is made from the same cork. I cut about 100 identical petals and arranged them in five rings on the floor. Then I painted them, using solid yellow for the inside ring and adding more and more blue as I moved outward. To mount the blossom, I ran concentric circular beads of adhesive and positioned each petal with a finishing nail—*Paul Levine. Drawing by Carl De Groote.*

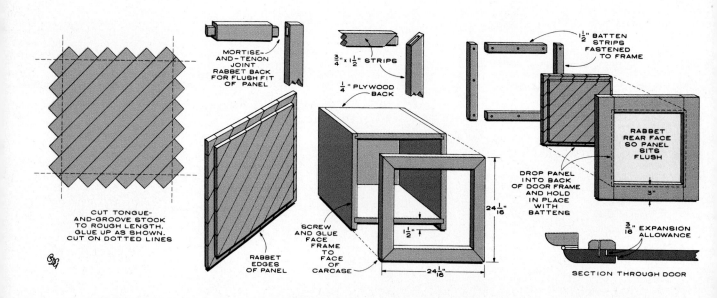

CUT TONGUE-AND-GROOVE STOCK TO ROUGH LENGTH. GLUE UP AS SHOWN. CUT ON DOTTED LINES

MORTISE-AND-TENON JOINT RABBET BACK FOR FLUSH FIT OF PANEL

RABBET EDGES OF PANEL

¾" x 1½" STRIPS

¼" PLYWOOD BACK

SCREW AND GLUE FACE FRAME TO FACE OF CARCASE

1½"

24 1/16"

24 1/16"

1½" BATTEN STRIPS FASTENED TO FRAME

RABBET REAR FACE SO PANEL SITS FLUSH

DROP PANEL INTO BACK OF DOOR FRAME AND HOLD IN PLACE WITH BATTENS

3"

³⁄₁₆" EXPANSION ALLOWANCE

SECTION THROUGH DOOR

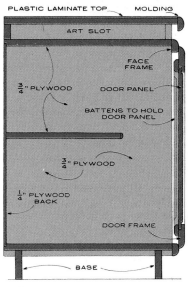

PLASTIC LAMINATE TOP — MOLDING

ART SLOT

FACE
FRAME

¾" PLYWOOD

DOOR PANEL

BATTENS TO HOLD
DOOR PANEL

¾" PLYWOOD

¼" PLYWOOD
BACK

DOOR FRAME

BASE

Carrie's inner sanctum combines practical storage with appealing design. Warm butternut, plus such decorative features as the free-form tree and cork flower, disguise this project's humble origins—a plywood box and particleboard shelves.

colonial cabinets

The colonial-style family room wall shown consists of a main cabinet, with two smaller units on each side. The size of your room will determine whether you build one or more modules. The main cabinet accommodates a TV and video recorder and provides ample space for books, games, and a display of your favorite keepsakes.

Earth-tone colors and the black, hammered hinges and locks on the raised-panel doors give the cabinets an authentic touch of early American charm. Despite its size, the project isn't difficult, and it requires only commonly available materials.

Four shelves are joined to the sides by dadoes, while the two remaining shelves are adjustable, supported by clips that mount into holes drilled into the sides. Face frames are assembled with mortise and tenon joints, but dowels may be used instead. The backs nest in rabbets machined into the rear inner edges of the sides, and nails lock them to all fixed shelves, integrating the construction. While you'd never guess it from the appearance, a simplified construction technique makes building the raised-panel doors easy.

First steps

Before starting, measure the length of your room wall exactly. Then, figure the width of the center cabinet—the one that will accommodate your TV set—and still allow an even number of 28-inch-wide cabinet modules on each side of it. Provide ½ inch for trimming and fitting the entire line of cabinets into the room length by using extra-wide, overhanging stiles on the end cabinets for scribing to the walls.

As the side cabinets are all the same, it's smart to cut out all the parts at one time, and then make up the center cabinet while the assemblies using these parts are drying. Begin by cutting out sides and shelves. Note that shelf No. 5, without edging, will be flush with the face frame when installed, while all others, except Nos. 1 and 6, include a pine edge in the overall dimensions for appearance. Shelves 1 and 6 don't require edging, as they'll be glued to the back of the face frame.

Glue the edges on shelves that require them, and sand them flush when dry. Rout the rabbets in the rear inner edges of all sides and lay out the shelf dadoes on the inner surfaces. You can make these dadoes several ways: (1) With a radial saw, set up dado blades to a width that's a touch over ¾ inch and cut the dadoes using table stops both for speed and accurate location. (2) Use a clamp guide and machine them with your router. (3) Score the dado outlines with a utility knife guided by a rafter square; then use your backsaw or portable saw with a veneer blade, guided by a fence to remove the waste.

Presand all exposed surfaces, test-fit the shelves in the dadoes, and, if all is well, glue and clamp or nail the basic cabinet carcasses together. Remove any excess glue immediately—while it's easy. Now, cut the backs and fasten them to the rabbets and shelves with glue and box nails.

Face frames are the next order of business. Each cabinet really has two—a top section with arches and a lower section into which the doors are inset. The top sections are made up of 1½-inch-wide stiles (vertical members) plus the arch pieces cut from plywood, while the lower section uses 1½-inch- and 2½-inch-wide stiles, a 2½-inch top rail, and a 3½-inch bottom rail. If you're not set up for ripping the stiles with a table saw, you can buy them from lumberyard stock as 1 × 2, 1 × 3, and 1 × 4.

Face frames

Before cutting face frames, decide whether you'll assemble them with mortise and tenon joints or dowels. Tenoned pieces require an additional ¾ inch on each end for machining.

Tenons can be cut on a table saw, with a router, or by hand. Mortises can be done on a drill press equipped with a mortising attachment; they can also be made by drilling a row of holes and chiseling away the waste, or completely by hand with a mortising chisel (available from Garret Wade Co., 161 Avenue of the Americas, New York, NY 10003, or Woodcraft Supply, 313 Montvale Avenue, Woburn, MA 01801). To make the arched pieces, set up your router with its trammel point and a straight-sided ½-inch diameter bit adjusted to swing a 4½-inch radius

arc. Lay out the arches on plywood, including the trammel point location. Then, starting with a minimal cut, swing the router back and forth, gradually increasing the plunge of the bit on each successive pass. When the arches are completely cut through, trim the sides, top and bottom to size. If you have no router, cut the arches with a jig or saber saw and sand.

If you're using dowels instead of mortises and tenons, dry-assemble the frames and clamp them together before drilling. Then, drill from the outside in, going completely through the

outer member. In either case, glue and clamp the frames together after fitting, and let them dry thoroughly.

Lay the cabinet carcasses on their backs and assemble the face frames to them with glue and nails. Line up the center stile of the upper section and nail it to the upper surface of shelf No. 5, making sure the face frames are flush against both its top and bottom surfaces. Then finish the carcass by setting the nails, filling, and sanding.

Cut the door stiles and rails, making the rails ¼-inch longer on each end to provide for the short tenons, unless

you're going to use dowels. Next, rout or saw out ¼ × ¼-inch grooves on all inner edges of the stiles and rails to receive the raised panels. Cut panel blanks from pine, allowing an extra ¼ inch all around to fit into the grooves you just made. Then, to prevent panel swelling from splitting door frames, shave ¹⁄₁₆ inch from all panel edges, so the panels "float" in their grooves.

Since these are bevel-raised panels, they can easily be machined on a radial saw with a Safe-T-Planer attachment (Wagner Co., Gilmore Pattern

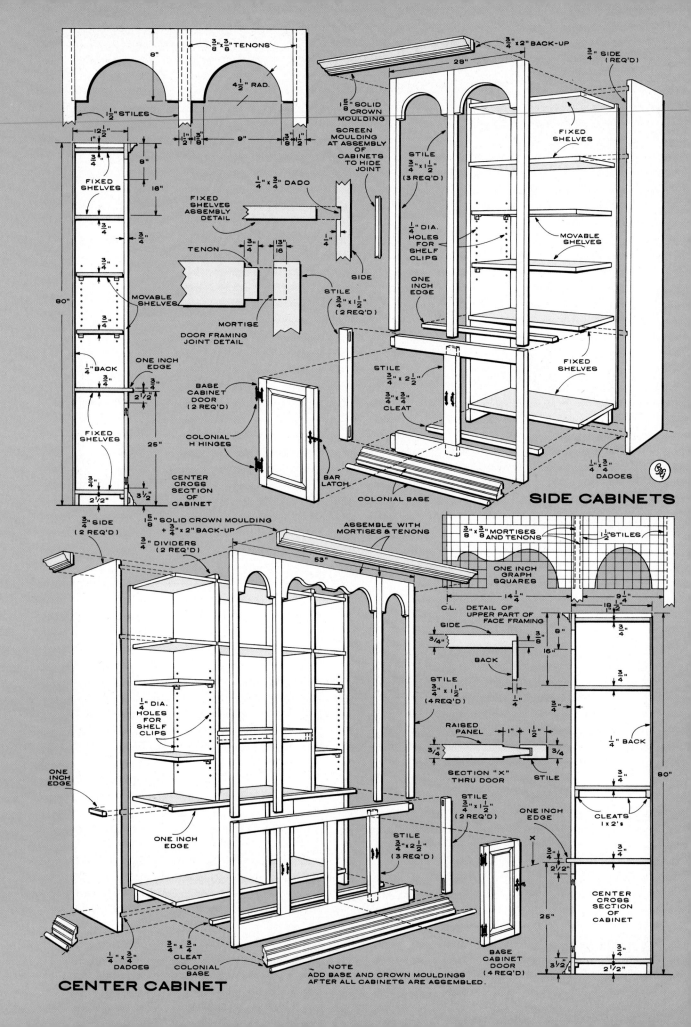

8"

$\frac{3}{8}$"x$\frac{3}{8}$" TENONS

4$\frac{1}{2}$" RAD.

1$\frac{1}{2}$" STILES

$\frac{1}{2}$" $\frac{3}{8}$"

9"

$\frac{3}{8}$" $\frac{1}{2}$"

12$\frac{1}{2}$"

1"

$\frac{3}{4}$"

6"

16"

FIXED SHELVES

$\frac{3}{4}$"

$\frac{3}{4}$"

$\frac{3}{4}$"

MOVABLE SHELVES

$\frac{3}{4}$"

80"

$\frac{1}{4}$" BACK

ONE INCH EDGE

$\frac{3}{4}$"

2$\frac{1}{2}$"

$\frac{3}{4}$"

FIXED SHELVES

25"

$\frac{3}{4}$"

2$\frac{1}{2}$"

3$\frac{1}{2}$"

CENTER CROSS SECTION OF CABINET

$\frac{3}{4}$" SIDE (2 REQ'D)

FIXED SHELVES ASSEMBLY DETAIL

$\frac{1}{4}$"x$\frac{3}{4}$" DADO

$\frac{3}{4}$" $\frac{13}{16}$"

$\frac{1}{4}$"

TENON

$\frac{1}{4}$"

SIDE

STILE $\frac{3}{4}$"x1$\frac{1}{2}$" (2 REQ'D)

MORTISE

DOOR FRAMING JOINT DETAIL

BASE CABINET DOOR (2 REQ'D)

COLONIAL H HINGES

BAR LATCH

SCREEN MOULDING AT ASSEMBLY OF CABINETS TO HIDE JOINT

1$\frac{5}{8}$" SOLID CROWN MOULDING

$\frac{3}{4}$"x2" BACK-UP

28"

$\frac{3}{4}$" SIDE (REQ'D)

STILE $\frac{3}{4}$"x1$\frac{1}{2}$" (3 REQ'D)

FIXED SHELVES

$\frac{1}{4}$" DIA. HOLES FOR SHELF CLIPS

ONE INCH EDGE

MOVABLE SHELVES

STILE $\frac{3}{4}$"x2$\frac{1}{2}$"

$\frac{3}{4}$"x$\frac{3}{4}$" CLEAT

FIXED SHELVES

$\frac{1}{4}$"x$\frac{3}{4}$" DADOES

COLONIAL BASE

SIDE CABINETS

$\frac{3}{4}$" SIDE (2 REQ'D)

1$\frac{5}{8}$" SOLID CROWN MOULDING + $\frac{3}{4}$"x2" BACK-UP

ASSEMBLE WITH MORTISES & TENONS

$\frac{3}{4}$" DIVIDERS (2 REQ'D)

53"

$\frac{3}{8}$"x$\frac{3}{8}$" MORTISES AND TENONS

1$\frac{1}{2}$" STILES

ONE INCH GRAPH SQUARES

14$\frac{1}{4}$"

9$\frac{1}{4}$"

18$\frac{1}{2}$"

$\frac{1}{4}$" DIA. HOLES FOR SHELF CLIPS

ONE INCH EDGE

ONE INCH EDGE

C.L. DETAIL OF UPPER PART OF FACE FRAMING

SIDE

$\frac{3}{4}$"

$\frac{3}{8}$"

8"

BACK

16"

STILE $\frac{3}{4}$"x1$\frac{1}{2}$" (4 REQ'D)

$\frac{1}{4}$"

RAISED PANEL

$\frac{3}{4}$"

1"

1$\frac{1}{2}$"

$\frac{3}{4}$

SECTION "X" THRU DOOR

STILE

STILE $\frac{3}{4}$"x1$\frac{1}{2}$" (2 REQ'D)

ONE INCH EDGE

STILE $\frac{3}{4}$"x2$\frac{1}{2}$" (3 REQ'D)

$\frac{3}{4}$" $\frac{3}{4}$" CLEAT COLONIAL BASE

$\frac{1}{4}$"x$\frac{3}{4}$" DADOES

BASE CABINET DOOR (4 REQ'D)

NOTE ADD BASE AND CROWN MOULDINGS AFTER ALL CABINETS ARE ASSEMBLED.

$\frac{3}{4}$"

$\frac{3}{4}$"

$\frac{3}{4}$"

$\frac{1}{4}$" BACK

$\frac{3}{4}$"

CLEATS 1x2's

$\frac{3}{4}$"

80"

$\frac{3}{4}$"

2$\frac{1}{2}$"

25"

$\frac{3}{4}$"

3$\frac{1}{2}$"

2$\frac{1}{2}$"

CENTER CROSS SECTION OF CABINET

x

CENTER CABINET

The easiest way to handle making of the cabinet arches is with a router and trammel point. The trick is to keep increasing the plunge with each pass.

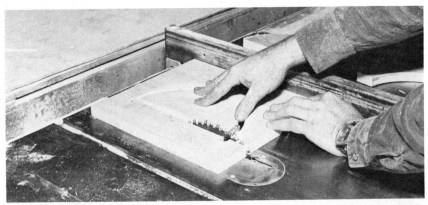

After you have machined out the arches, trim the tops, sides, and bottoms with the table saw and then fine-sand all of the edges.

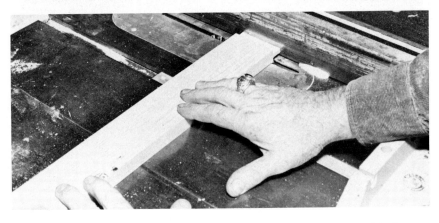

The quickest way to make the tenons in the home workshop is on the table saw. First step is to cut the shoulder lines. After that, you are ready to crossfeed.

Works, 1164-R North Utica, Tulsa, OK 74110), or on a table saw by angling the blade.

When you assemble the doors, use no glue between the panel edges and the rail and stile grooves, only at the corner joints. Make sure the doors are square when you clamp them up. Fit the inset doors into the frame openings using a block plane (which can be operated with one hand) and identify

Materials
Paint: Martin Senour, Williamsburg, John Greenhow Green, and Coachhouse Green.
Hardware: Latches—Amerock No. 8572; Mc-Kinney No. 537. Hinges—Amerock No. 1616; McKinney No. 520. Shelf clips—Amerock No. 3828.

both doors and openings. It's best to provide at least ¹⁄₁₆-inch gaps all around between the door and the frame to avoid sticking later. Now, mount the hinges and latches and make any final corrections.

Make a drill template from scrap ¼-inch plywood and use it to drill all the shelf-support clip holes (¼-inch diameter) on the inner faces of the sides.

Construction of the center cabinet is quite similar to the smaller ones, except the shelf that holds the TV is cleated and reinforced, and the arch shapes must be cut with a pattern or by hand.

Final steps

Take off the hardware and check that all surfaces are filled and sanded. Re-

move all dust, first by vacuuming, and then by wiping with a tack cloth. Apply a couple of coats of enamel underbody, sanding after each coat. Remove all dust and give everything a finish coat of paint. When dry, reinstall all H hinges, bar latches, and shelf clips.

To install the cabinets, first locate the wall studs and fit the cabinets temporarily in place. Scribe the outer stiles to the walls as necessary, and mark the stud locations on the cabinets. Line the cabinets up horizontally by placing shims under any low spot, and vertically, by placing shims between the wall and the cabinets at the stud locations. Add base, crown, and screen moldings (to hide joints between cabinets), and paint them—*Bernard Price.*

three
elegant carts

Americans are famous for eating on the run. And we often find ourselves moving around in our homes to eat: barbecues and breakfasts outdoors in warm weather, snacks around the television, or a meal out on the deck as a break from the table routine. When we do change scenes, something is usually left behind—the knives, the coffee, or the ketchup—and we jump up and trot back to the kitchen fetching the missing item.

An answer to that problem is one of these handy cart designs from Georgia Pacific (see photos and drawings). You can either load a cart in the kitchen and roll it to where you're eating, or you can plant a well-stocked cart where you often prepare food or entertain. The appliance cart brings cooking tools back to the kitchen, too.

Best of all, the three carts are planned for construction with common sizes of lumber: 1×2s, 2×2s, and so on. You don't need a shopful of power tools to recut and dress the wood. (A portable circular saw will be a big labor saver when cutting large plywood sections, and a sabre saw is a must for the free-form cutouts on the serving-cart frames. Otherwise, hand tools are all you'll need.)

A two-tone paint job brightens up the carts' appearance, or you may want to finish boards with a clear, exterior sealer.

More detailed plans and patterns for the serving cart and appliance center are available in a booklet that includes eight other projects. Write to Redi-Cuts, PS, Georgia Pacific, 900 S.W. Fifth Avenue, Portland, OR 97204. For more details on the barbecue cart, write to Housekit, Georgia Pacific, same address—*Paul Bolon.*

Serving Cart

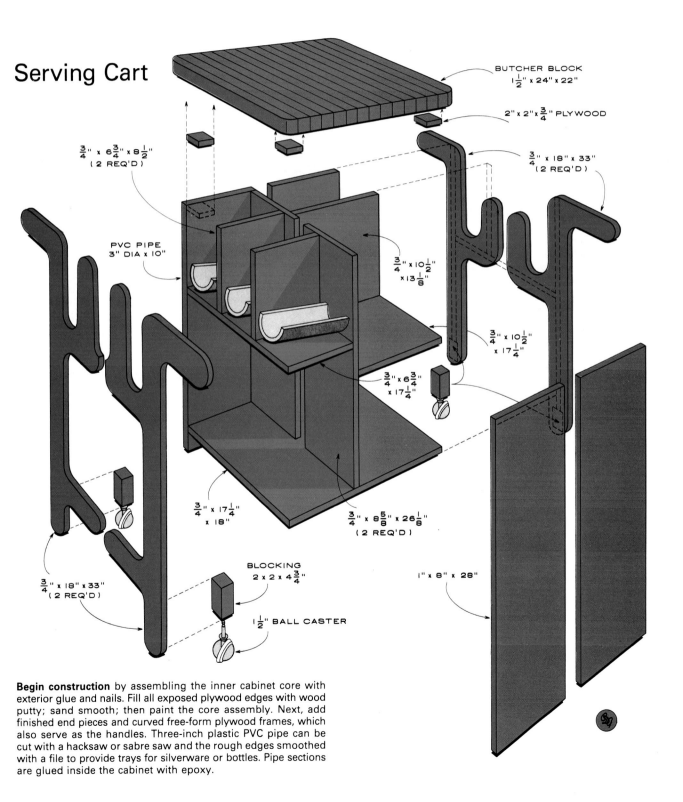

BUTCHER BLOCK
$1\frac{1}{2}"$ x 24" x 22"

$2"$ x $2"$ x $\frac{3}{4}"$ PLYWOOD

$\frac{3}{4}"$ x $6\frac{3}{4}"$ x $8\frac{1}{2}"$
(2 REQ'D)

$\frac{3}{4}"$ x 18" x 33"
(2 REQ'D)

PVC PIPE
3" DIA x 10"

$\frac{3}{4}"$ x $10\frac{1}{2}"$
x $13\frac{1}{8}"$

$\frac{3}{4}"$ x $10\frac{1}{2}"$
x $17\frac{1}{4}"$

$\frac{3}{4}"$ x $6\frac{3}{4}"$
x $17\frac{1}{4}"$

$\frac{3}{4}"$ x $17\frac{1}{4}"$
x 18"

$\frac{3}{4}"$ x $8\frac{5}{8}"$ x $26\frac{1}{8}"$
(2 REQ'D)

BLOCKING
2 x 2 x $4\frac{3}{4}"$

1" x 8" x 28"

$\frac{3}{4}"$ x 18" x 33"
(2 REQ'D)

$1\frac{1}{2}"$ BALL CASTER

Begin construction by assembling the inner cabinet core with exterior glue and nails. Fill all exposed plywood edges with wood putty; sand smooth; then paint the core assembly. Next, add finished end pieces and curved free-form plywood frames, which also serve as the handles. Three-inch plastic PVC pipe can be cut with a hacksaw or sabre saw and the rough edges smoothed with a file to provide trays for silverware or bottles. Pipe sections are glued inside the cabinet with epoxy.

Barbecue Cart

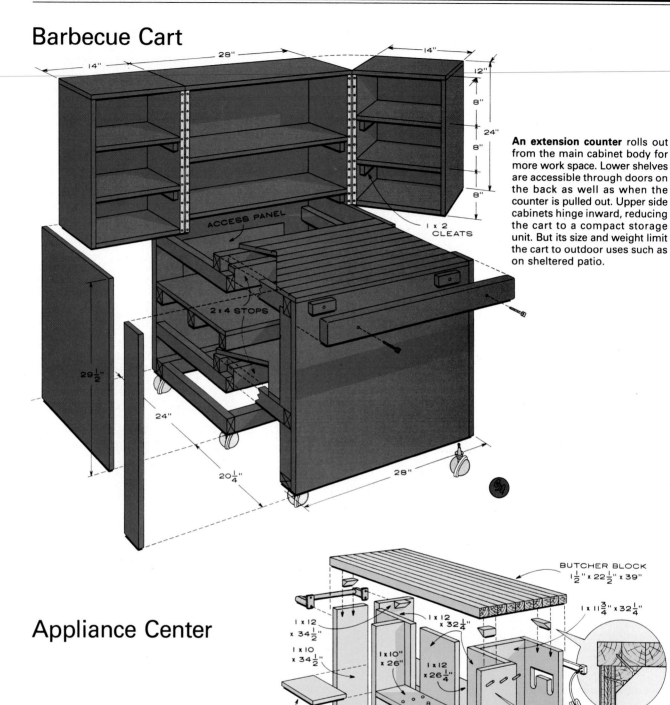

An extension counter rolls out from the main cabinet body for more work space. Lower shelves are accessible through doors on the back as well as when the counter is pulled out. Upper side cabinets hinge inward, reducing the cart to a compact storage unit. But its size and weight limit the cart to outdoor uses such as on sheltered patio.

Appliance Center

This setup is designed as a work area to accommodate all the portable appliances in a kitchen. Cutout on the back panel is for the cord of a 10-inch power strip with continuous outlets. For the best finish, the cart should have two or three coats of exterior paint or sealer. Ideal paint job is one coat of an oil-base paint followed by two coats of enamel. The plan here is modified from a detailed plan in the Redi-Cuts booklet.

inside shutters that save energy

If your house is well insulated, you can double-glaze, triple-glaze, and weatherstrip your windows until you are sick of the whole business, and the windows will still be your biggest source of heat loss. The ad of a leading window company humorously suggests that the only cure is to board up the windows.

A less drastic solution is well-fitted, weatherstripped, wood shutters, mounted inside the windows. While insulated wood shutters are more effective than plain wood ones, either is more effective than triple-glazing. Triple-glazing in the sash can reduce heat loss over double-glazing by 18 percent; well-fitted, weatherstripped, wood shutters will save 48 percent, and insulated shutters 61 percent over double-glazing.

Heat is lost through a window two ways. Heat passing through the glass is called *transmission loss*. It can be reduced by trapping a layer of dead air between two layers of glass. It can be further reduced by trapping a second layer of dead air by triple-glazing, or by adding a shutter. A shutter is better than glass, as the wood itself blocks the flow of heat.

Infiltration is the second form of window heat loss. It is air passing through the cracks between the sashes and the window frame. Weatherstripping and tight-fitting window sashes reduce infiltration. A well-fitted shutter with weatherstripping can cut this loss by providing an additional barrier.

Energy-saving capability is just one advantage that wooden, interior shutters have over triple-glazing. Privacy is another—shutters effectively seal off a room from the outside. Shut-ters may even contribute to your home's security; if a potential intruder can't see in, and there's obviously more there than a window shade or draperies, he's likely to play it safe and go elsewhere.

On the following pages are three approaches to building interior shutters. But these designs are only a starting point.

For one thing, the shutters shown are not likely to fit your windows, so you'll have to modify the plans to suit your needs. Here are other things to consider in designing your own energy-saving shutters:

● You may want to substitute ⅛-inch hardboard for plywood in insulated shutters. This will reduce the thickness to 1 inch, with little reduction in the desired thermal insulation.

● Allow ¹⁄₁₆-inch clearance all around, and between shutters to avoid sticking. Also, do not leave more than ¹⁄₁₆ inch between the back of the shutter and the wood doorstop.

● While you would expect closed-cell foam weatherstripping to be more effective than the lower-priced, open cell type, we found that the open-cell foam stayed in place better (though neither will last forever).

● For a tight fit, shutters must be latched to the window frame, not to each other.

You'll find that you can build energy-saving shutters for less than you would have to pay for light-weight interior shutters. Your biggest outlay will be for hinges and latches.

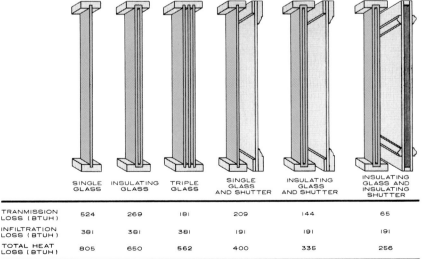

	SINGLE GLASS	INSULATING GLASS	TRIPLE GLASS	SINGLE GLASS AND SHUTTER	INSULATING GLASS AND SHUTTER	INSULATING GLASS AND INSULATING SHUTTER
TRANMISSION LOSS (BTUH)	524	269	181	209	144	65
INFILTRATION LOSS (BTUH)	381	381	381	191	191	191
TOTAL HEAT LOSS (BTUH)	805	650	562	400	335	256

INFILTRATION LOSS CALCULATED FOR ALUMINUM OR WOOD DOUBLE-HUNG WINDOWS

Effectiveness of shutters as energy-savers is shown below. Btuh losses are given for 8-square-foot windows, 58-degree inside-outside temperature difference, 15 mph wind. Window sashes are average fit, and like shutters, weatherstripped.

Design 1

These shutters were designed for a basement powder room, facing north, that was always colder in the winter than other rooms. Both sides of the shutters are decorated, making them equally attractive open or closed. The shutters are lapped for an air-tight fit.

Construction: Cut plywood panels to slightly oversize dimension, cut interior frame parts to size, miter decorative molding. Assemble mitered moldings and nail and glue to plywood, running the nails through the plywood, into the molding. Assemble the frame with glue only, then glue and nail to one face panel. Keep the nails out of hinge, lapped edge, and latch areas. Cut polyfoam insulation to fit, insert, glue, and nail second panel. Rout the mating edges to form the overlap. No-mortise hinges were used.

Shutters fashioned for basement powder room are shown at top, right. You can miter decorative molding for just one, or both sides of the shutters (above). Although moldings can be glued and nailed to shutter face panel one by one, it is easier and faster to assemble the decorative frame in a frame clamp, or set of corner clamps (center). Before applying glue, position decorative frame on plywood panel as shown above, right, clamp, turn over, and pin into position with partially driven nails. After gluing, drive nails all the way and clamp. Once an internal frame has been glued to a shutter panel, cut polyfoam insulation to fit and glue it to the second side, as shown in the photo below, right.

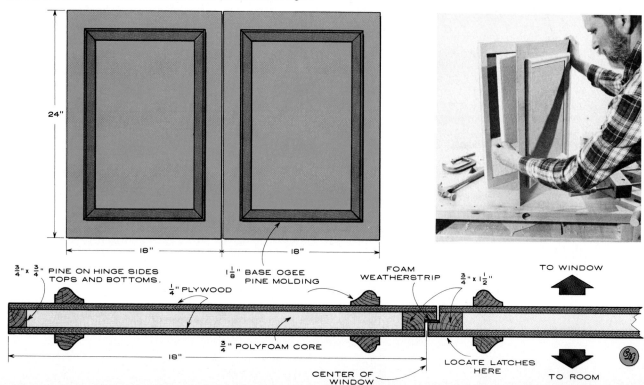

24"

18"   18"

¾" × ¾" PINE ON HINGE SIDES TOPS AND BOTTOMS.

¼" PLYWOOD

1⅛" BASE OGEE PINE MOLDING

FOAM WEATHERSTRIP

¾" × 1½"

TO WINDOW

¾" POLYFOAM CORE

18"

CENTER OF WINDOW

LOCATE LATCHES HERE

TO ROOM

Design 2

These uninsulated wood shutters are mounted on loose-pin hinges so they can be taken down during the warm part of the year. When up, they are kept closed most of the time, so only one side of the shutters was decorated.

Construction: Blank rails, lay out curved edge, saber-saw to outline, and sand smooth. Cut posts overlength for ease of clamping when routing. Glue frames together, clamp to bench (spaced up with ¾-inch scrap), and rout opening with ball-bearing piloted Roman ogee bit. Glue ¼-inch plywood panels to frame backs and trim panel backs with lattice. Mortise for hinges and rout slots for tongue-and-groove closure. Glue foam weatherstrip in groove for snug fit when closed.

(Continued)

These shutters are intended for easy removal. Surface bolts secure them to window frame and hold them tightly against weatherstripped molding. Above: Curved side of rails are cut with saber saw. Below: After cutting, rough edge is smoothed with drum sander.

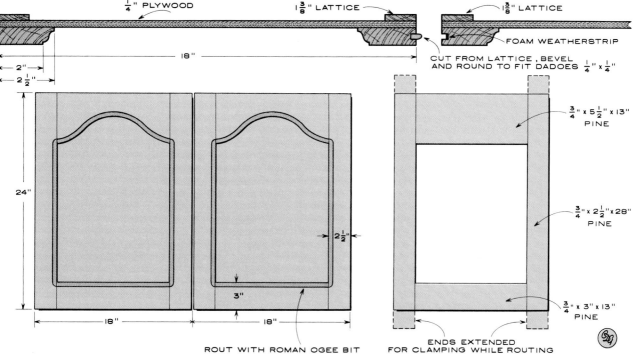

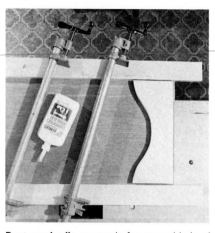  

Posts and rails are ready for assembly in photo at left. Glued butt joints were reinforced when the panel was glued on. Projections are for clamping while routing. Center: After gluing plywood panel and lattice backside trim to frames, slots are routed in mating edges for tongue-and-groove closure. Bottom of groove is lined with foam weatherstrip. Tongue is ripped from lattice; profile requires additional rounding for smooth closing. Right: As hinge mortises extend across full thickness of shutters, both shutters are clamped and routed back to back. Simple clamp-on jig guides router.

Design 3

Poorly weatherstripped tracks allowed cold air to get through our 6-foot wide, sliding aluminum windows, even with storms. To cover the windows, we made shutters in folding pairs. The shutters are oak (it was cheaper than available select pine). Eight surface bolts hold them tightly against the weatherstripped doorstop, which minimizes infiltration.

Construction: Cut frames to dimension, allowing for rail tenons. Rout slots for plywood panels; then rout post ends to form mortises. Saw tenons to fit mortises. Glue three sides of each frame; then insert panel and glue fourth side. Trim shutters to fit window and rout hinge mortises. We hinged our shutters to the window frame with 3 × 3 butts; between pairs, we used invisible hinges, but ordinary butts would do. Facing sides of shutters have shallow tongue and grooves. The round-bottom groove is routed with a ½-inch core bit; the mating tongue is ½-inch, half-round molding, set in a ¹⁄₁₆-inch-deep dado. As with other shutters, foam weatherstrip was glued in mating groove for airtight fit. Rout mortises for hinges, and slots for tongue and groove—*Thomas H. Jones.*

Shutters designed for 6-foot-wide windows. In summer, they help keep room cool.

Sawing tenons on shutter rails can be done accurately in miter box (top), if stop-block is clamped to box and bench to keep tenon lengths identical. Tenon is finished with chisel. Shutter frames and panels are assembled with glue (bottom).

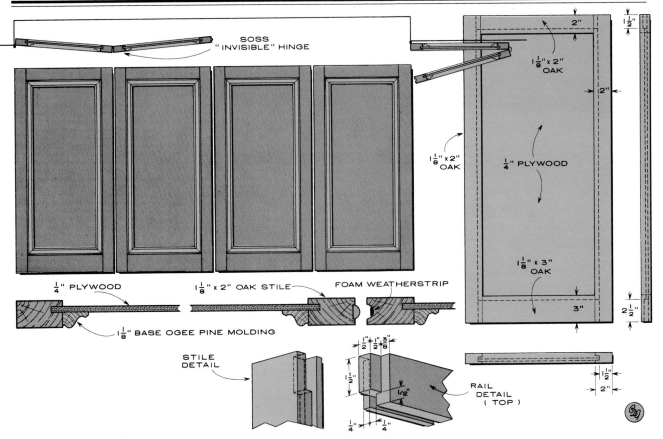

SOSS "INVISIBLE" HINGE

2"

1 1/2"

1 1/8" x 2" OAK

2"

1/4" PLYWOOD

1 1/8" x 2" OAK

1 1/8" x 3" OAK

3"

2 1/2"

1/4" PLYWOOD

1 1/8" x 2" OAK STILE

FOAM WEATHERSTRIP

1 1/8" BASE OGEE PINE MOLDING

STILE DETAIL

RAIL DETAIL (TOP)

1/2" 1/2" 3/8"

1 1/2"

1/2"

1/4" 1/4"

1 1/2"

2"

Construction details for Shutter 3. Top row: Left—Hinges are 3×3, loose-pin butts, set in mortises. Right—Soss invisible hinges join shutter pairs. Middle row: Left—Doorstop molding is positioned about 1/16 inch behind shutter. Center—Adhesive-backed, foam weatherstrip on doorstop is tightly compressed when shutter is latched. Right—Surface bolts latch shutters to sill. Bottom row: Left—Brass-plated, steel-type surface bolts (left in photo) are available at Sears; solid-brass type (right), at hardwares. Center—Strike plate for bolt is mortised into sill. Right—Frames were stained.

floating stairs

Louise Rigg's stairs are a flight of fancy. The dramatic steps, which hang from the soaring ceiling of Rigg's modern Berkeley, California, home, give one a sensation of floating, the interior designer says. "It's an unattached feeling, not like climbing at all. But people who are nervous about heights must be careful."

The steps are not Rigg's first floating staircase. Most of the thirteen homes she has built and lived in over the years have included a variation on the theme, but his collaboration with architect Alex Achimore may be the most successful. The design is not just a thing of beauty; it also opens more usable floor space in the redwood-paneled living room—which Rigg uses to striking effect as the setting for her grand piano.

The steps, laminated assemblies of 2-inch-thick planks, are supported on both ends: They are tied into the adjacent wall with hidden steel-plate angle brackets, and they hang from the ceiling on threaded steel rods. Achimore said that the support rods could have been eliminated and the steps fully cantilevered, but a stronger wall would have been necessary. In fact, Rigg did just that with her previous stair design, but then the supports ran through the wall into an adjoining garage. In this case the wall faced an outdoor patio, so the structure had to be entirely contained within the wall. Besides, the support rods serve a second purpose as hand grips. Copper sleeves around them blend visually with the redwood theme.

The wall is a double-stud frame with a 3 × 12 stringer running from the lower to the upper level. Support

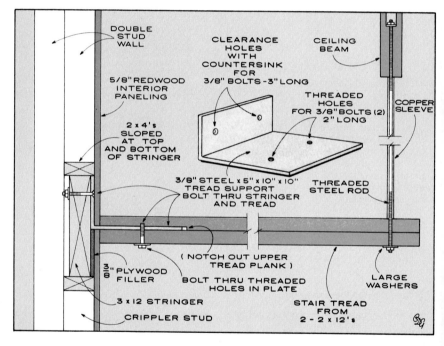

brackets fabricated from ⅜-inch-thick steel plate are bolted through the stringer and into a groove notched in the stair treads. The stringer is sandwiched by parallel 2 × 4s that hold the space needed for the brackets, and additional studs run to ceiling and floor as backing for panel treatment.

The vertical supports are anchored with nuts and washers into a beefed-up roof beam. Since that beam had to be drilled and installed before the roof went on (long before the stairs were installed), it created some interesting problems for the builder. According to Achimore, the contractor laid out the holes in the beam by placing the stairs on the floor and shining a vertical light from a homemade laser onto the

overhead support. Then he drilled holes and secured the steel rods before proceeding with the roof. The rods dangled there for more than a month until finish work was begun and the stairs were hung.

The entire house—inside and out— is covered with redwood, partly because of its resistance to fire and termites and its good acoustical properties, but mainly because its warm texture complements the owner's taste for unusual shapes and angles. The material is not strong enough to support great weight, however, so both the stair treads and ceiling beam are made from Douglas fir, stained to match the vertical-grain paneling— *Daniel Ruby. Photos by Jeff Weissman.*

dishwasher island console

These unique installations offer not only a place for a dishwasher in a cramped kitchen, but generous storage space as well. And what better place to stow dishes, pots, pans, and tableware but right next to the machine that keeps them clean?

● The island, intended as home for a portable dishwasher, has generous double-access storage behind the machine.

● The peninsula, made for a built-in dishwasher unit, offers a clever rolling cart that can travel from kitchen to dining table, or out to the patio when you're picnicking. Slipped into its spot next to the dishwasher, it blends with existing cabinets.

In the kitchen/dining area shown, the peninsula takes up the space of one regular base-cabinet unit. At right angles to the existing counter, it's located near the sink. It also divides the kitchen work area from the adjacent dining area.

The island takes advantage of this portable dishwasher's cutting-board top. You can locate it anywhere there's floor space: centered in the kitchen work area, or even positioned against a wall. (In the latter case, you'd eliminate one set of cabinet doors.) The unit's fold-down shelf gives you a quick spot for snacking, menu planning, or food preparation.

The drawings below give detailed dimensions and show the materials needed. Note that dimensions shown will accommodate Maytag's model WU601 built-in dishwasher (for the peninsula) and model WC401 portable (for the island). Alter the dimensions to suit other makes and models.

Want further information? Installing the built-in dishwasher will be easier if you first read a thirty-six page booklet that's 25 cents from Maytag (Newton, IA 50208). It describes utility requirements, how to provide drainage, and how to cut and fit tubing—*Richard Stepler*.

Island location needs no plumbing hookup: Dishwasher is portable and wheels to sink for connection. Island also offers double-access storage cabinet, plus a sturdy fold-out shelf for extra serving space.

Built-in dishwasher's peninsula addition to an existing counter (left) also holds a handy pullout rolling cart (shown above) for convenient, portable storage of dishes, other tableware.

HOW TO BUILD THE ISLAND

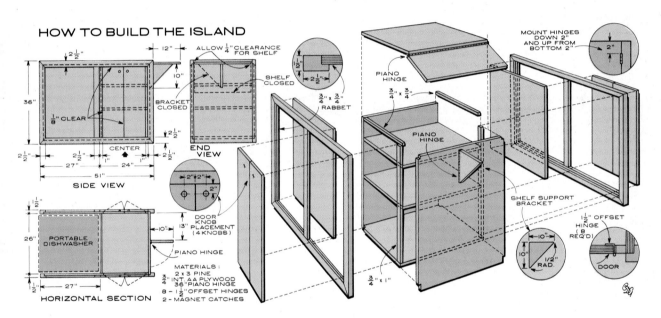

HOW TO BUILD THE PENINSULA

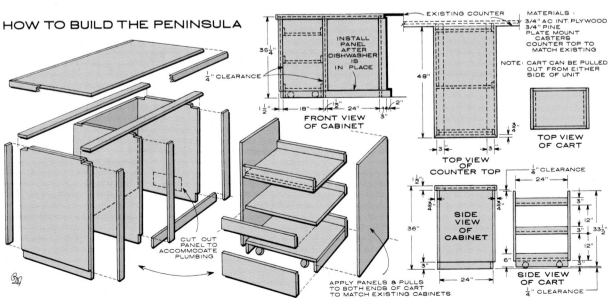

bed shelving

Here's a bedroom where a bed/shelf unit becomes the focal point, encompassing an entire wall and using otherwise wasted space.

The bed underframing is a rectangular box (think sandbox) made of mitre-cut, kiln-dried 2 × 10-inch lumber. The mattress foundation consists of two slabs of 1⅛-inch plywood. Two 4 × 8-foot sheets of top-grade, tongue-and-groove flooring will do for this. For a queen-size mattress, cut two 40 × 60-inch slabs; for a standard, double mattress, two 37½ × 54-inch slabs; for a king size, 40 × 77-inch slabs. Sand rough edges and slightly round corners of the slabs with medium-grit paper. Edges can be trimmed with 1⅛-inch bull nose stop, and the slabs then painted to match

the frame and should be painted if bedding will be tucked. Or you can do as I did and leave this step for a later date, since the slabs will be concealed by bedding.

The shelves that span the wall above the bed head are built of 2 × 10-inch kiln-dried lumber. Avoid green lumber—it may warp, split, and shrink as it dries, and moisture sealed into the wood by paint could cause the paint to crack and chip. The shelf unit is assembled without glue, nails, screws, or any other fasteners. Vertical uprights are held fast to the ceiling with adjustable tension devices, and horizontal shelves merely slip into dado grooves.

Start by trimming the four vertical uprights for shelving. If your bedroom has a standard 8-foot ceiling and is

carpeted, the boards will be 94 or 94½ inches, depending on floor-covering thickness; with uncarpeted floors, they'll be 95 inches.

Now lay out all four uprights and, starting at the bottom, scribe lines across them at 4, 5½, 17½, 19, 34, 35½, 47½, 49, 64, 65½, 77½, and 79 inches. You will notice that these are pairs of lines 1½ inches apart that are separated alternately, by 12 and 15 inches.

If you don't own a router and can't borrow one, you might rent one. (A word of caution: If you've never used a router, be sure to practice on scraps, and always clamp any material you rout to a solid work surface.) Use the router to make ½-inch-deep, 1½-inch-wide dado grooves between the paired lines. To assure accuracy, make a router guide from wood scraps, or

clamp a straightedge across the board in front of the router to keep the tool on a true, left-to-right track. Check each dado with a piece of 2 × 10 to see that shelves will fit snugly into the grooves. Use a narrow rasp and coarse sandpapar to remove just enough wood in the grooves for a proper fit.

Once all dadoes are routed, two uprights—they will be the extreme left and right vertical ones—are finished. The other two, the center uprights, will need two dado grooves to accommodate the center shelves. Lay these uprights out, grooved sides down, and scribe lines across at 57, 58½, 72½, and 74 inches from the bottom. Rout grooves between the paired lines as before.

Use a drill guide and ½-inch bit to make a starter hole 2 inches inside the front and rear edges of each vertical upright. Then use a wood bit to drill holes to size for the tension devices. Push female portions of the devices into holes and gently tap them in with a hammer.

Shelves are cut shorter for the sides of the unit, longer for the center. Two variables determine the cutting measurements: length of the wall the unit will span and bed width. There should be 4-inch clearance on each side of the bed from the center vertical uprights. For a queen-size bed, center shelves should be about 69 inches long; for a double bed, about 63 inches; and for a king, 86 inches. But wait to cut these shelves; lengths might vary by a fraction of an inch and they must be cut precisely to fit.

First, cut the twelve side shelves. Our bed, a queen size, called for 30-inch side shelves. Had it been a

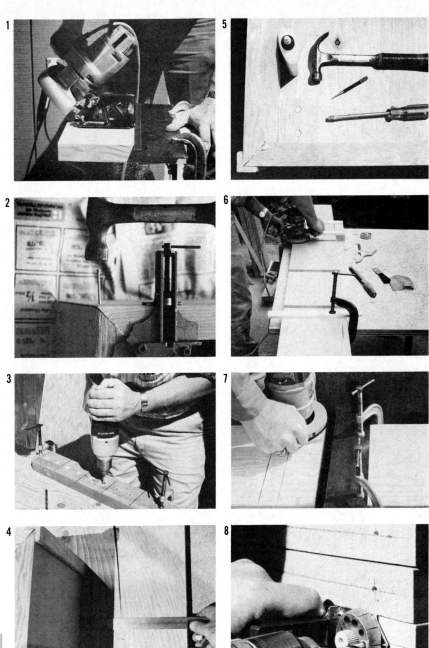

Start (1) by miter-cutting two 2 × 10 bed-frame panels to 60 inches, two to 40 inches (queen-size). Apply glue to ends, join to form box. **(2)** Using miter clamp to hold frame joint, secure with four 8d finishing nails—two from each direction. With #120 paper, sand edges. **(3)** From scrap 2 × 4, make four 4-inch stop-blocks. Drill and countersink two ¼-inch holes in each. **(4)** Center frame on plywood slabs, set face down. **(5)** With glue and 2-inch #12 flathead wood screws, attach stop-blocks to slab undersides at each inside corner of frame. Start wall unit by **(6)** clamping 2 × 10 vertical upright to solid surface and **(7)** carefully measuring, marking, and routing grooves for shelves. **(8)** Use drill guide and ¼-inch bit for starter hole at top of each upright.

standard double, shelves would have been 33 inches to compensate for a 6-inch narrower bed; had the wall been 14 instead of 11 feet, with a queen-size bed, the side shelves would have measured 48 inches—you get the idea. Once the side shelves are cut, sand

them and the uprights lightly, wipe them clean, and paint to suit the room decor.

Start shelf assembly by standing the extreme left and right uprights in position at the corners of the room. Protect the ceiling from marks by placing

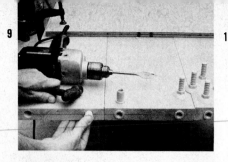

To continue work on wall section: **(9)** Use wood bit to drill holes for tension devices. Insert female section into holes and tap in gently. **(10)** Cut twelve side shelves from 2 × 10 stock. (Note the homemade saw and router guide used to assure accuracy.) **(11)** Begin assembly by standing vertical uprights against wall and securing to ceiling with tension devices. **(12)** Measure for two center shelves; then cut, sand, and paint them. Slide the shelves into place. **(13)** The space under the plywood slabs, inside the bed frame, provides storage space for bedding and other items. **(14)** The fluorescent lights under the center wall cabinet make good reading lamps and are simple to install.

circles cut from a new household sponge between it and the tension devices. Tighten the devices with a wrench.

Stand the center uprights in place so the end uprights brace the side shelves. Slide the shelves into the dado grooves, adjusting the center uprights as required. When all shelves are in place, make final adjustments and tighten the tension devices on the center uprights.

Now is the time to measure for those center shelves mentioned earlier. Stretch a steel tape measure from the inside of one dado groove to the other, and double check the distance. Cut two shelves to length, paint them, and slide them into their grooves to finish the wall unit.

Customizing

There are several ways to modify the shelf unit to fit your own needs and tastes. For example, although the center shelves need no additional support, a small vertical upright can be cut to 15 inches and inserted into dado grooves that have been routed into the center of these shelves, dividing them into separate shelves. I added two such vertical uprights and created 11½-inch-wide cubbyholes on each end of the center shelves, where we keep our nightcaps—not the cranial variety, but a decanter of brandy and two snifters on one side, a decanter of sherry and glasses on the other.

Another improvement I made later was the addition of four 9½-inch pieces of 1 × 2 that I inserted between the tension devices and the ceiling. This simple modification helped give the unit a permanent, built-in appearance.

Lighting fixtures can be attached to the wall above the bed, to the sides of the center uprights, or to the bottom of the lower center shelf. We decided on the latter and used inexpensive, fluorescent under-cabinet lights that are easy to install.

Alternate Shelf Construction Methods

Design of this bedroom shelf unit can be modified to fit needs. The distance between shelves can be altered, and the assembly method can be changed entirely.

If the idea of dadoed grooves doesn't appeal to you, you might wish to butt the shelves to the uprights and attach them with wood screws. For this method use 3½-inch #12 wood screws in each end of the side shelves. The center shelves should have the added support of a shelf bracket attached to the wall on the underside of the lower shelf.

A simpler modification calls for using cabinet-type metal shelf supports. These are small, L-shaped fixtures that fit into holes drilled into the vertical uprights. If you want adjustable shelves, drill two lines of holes, spaced 3 to 6 inches, in each vertical upright. Then insert the metal shelf supports. Or, instead of cabinet-type shelf supports, you can make your own out of ½-inch dowel rod. If you decide against the dadoed shelf joints, remember to cut your shelves an inch shorter than those in my original design.

Perhaps you want a more permanent unit. Then you can forget the tension devices. Instead, affix the extreme left and right verticals to the walls with screws. Center uprights can then be attached to the top and bottom side shelves with wood screws. The center shelves and remaining side shelves can be slipped in place as in the original design. Now you have a semipermanent unit that, with only slight effort, can be dismantled for moving, redecorating, painting.

The design of this unit is flexible enough to accommodate just about any size bedroom. Its mobility is particularly suited to the renter who will want to take it with him when he moves or to the family that might wish to put a child's bedroom to some other use when the youngster grows up and moves out.

The shelf unit goes well in any room where shelves are needed. A desk work table, sewing machine, or love seat would fit nicely between the center uprights. So if the shelves outlive their usefulness in the bedroom, you can move them to another room—*Kenn Oberrecht.*

bed under floor

Remove a bed and the "found" space in a room amounts to what you see in this boy's room. A trundle bed is an old space-making trick useful for storing a guest bed anywhere. And this bi-level arrangement, covered with Congoleum sheet vinyl, is a good way to make use of the space over it. Base molding is removed for platform construction, replaced around top.

Materials: 2×4 plates—one 10 feet long, two 2 feet long; 2×10 beams—four 6 feet long; 2×4 joists—seven 10 feet long; 2×4 bed framing—two 75 inches long, two 42 inches long, two 39 inches long; ½-inch plywood flooring (sub) and ½-inch plywood bed board—three 4×8-foot sheets; ¾-inch plywood—16 inches × 10 feet; blocking (bed) 2×6—1 foot long; ¼×1½-inch molding—10 feet long; heavy-duty casters—four 2- or 2½-inch; finish flooring of your preference, with the appropriate cement for application; three drawer pulls in any style you choose; a good supply of nails.

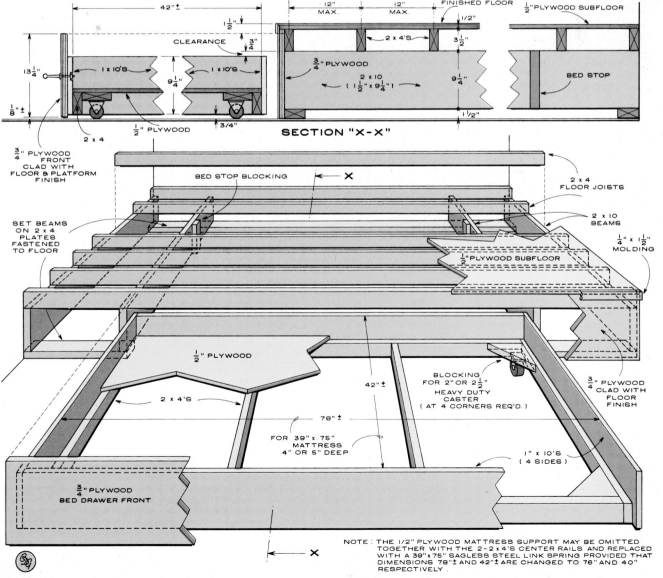

SECTION "X-X"

NOTE: THE 1/2" PLYWOOD MATTRESS SUPPORT MAY BE OMITTED TOGETHER WITH THE 2-2×4'S CENTER RAILS AND REPLACED WITH A 39"×75" SAGLESS STEEL LINK SPRING PROVIDED THAT DIMENSIONS 78"± AND 42"± ARE CHANGED TO 76" AND 40" RESPECTIVELY.

swing-out partition shelves

If your house has a pair of Lally columns supporting an upper floor, you can use them to anchor a shelved storage cabinet with a set of swing-out doors. The doors—with adjustable shelves—give you an easy-access, roomy partition unit. Face it with perforated hardboard, and you can use Peg-Board hooks and brackets to hang small items from front and back.

To secure the unit, fasten 2 × 8s to the support columns with U-bolts. I made mine by heating threaded ⅜-inch-diameter rods with a propane torch to bend them to shape, using the columns as a form. Mark the supports, drill two holes for each rod, then fasten with flat washers and nuts. Use three rods for each side.

Cut the shelving for the stationary section to length. Support each shelf at both ends with uprights of the same stock cut to the desired height. Starting at the bottom, wedge in the shelving supports and shelves successively, supporting the shelves in the center with pieces of the same height nailed through each shelf as it is positioned. On the back, attach sheets of ⅛-inch Peg-Board.

To make the doors, use ¾-inch plywood cut into 12-inch strips for each unit's sides, top, and bottom. Back this frame with Peg-Board and mount a 3-inch-diameter wheel on the bottom of each unit. Shelves in the doors can be made fixed or adjustable using any suitable hardware.

Attach the door units with 6-inch strap hinges attached at three points as shown—*W. David Houser.*

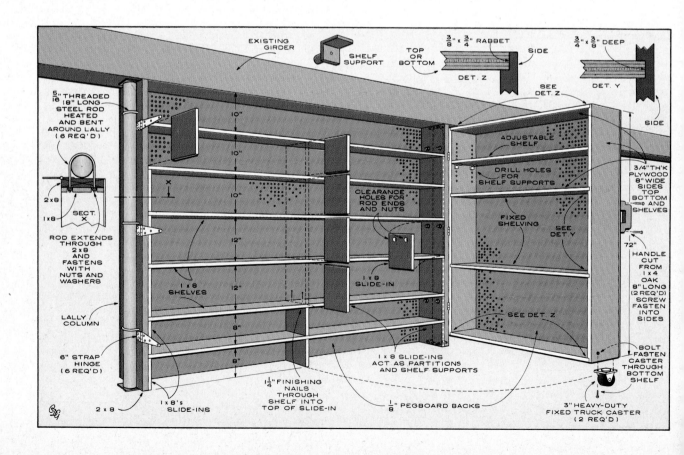

Doors open easily (above) on wheels secured to bottom shelf. Below, Peg-Board on back of the unit is for hanging tools.

laundry storage center

Even at best, a homemaker's close encounters with the basement laundry room are not likely to be the high point in home fun. If the room happens to be dismal and inefficient, those encounters can be awful. The solution: Make your laundry set up as attractive and efficient as possible.

The conversion shown here provides cabinet storage for all cleaning necessities, a folding area for finished laundry, and a rolling hamper cabinet. Rich oak panels have been installed over a drywall-reinforced stud wall, and electrical outlets, vents, and service lines have been neatly positioned for accessibility and appearance.

All cabinets and appliances fit flush against the wall and the room has a suspended ceiling and vinyl tile floor. This project, well within the ability of the average do-it-yourselfer, takes only about a week to complete, but will provide years of satisfaction.

Before starting, decide on the exact location and gather up all the required materials. Begin the job by erecting the stud wall(s); do it right and the rest is downhill.

The wall structure (studs, top and bottom plates) must be plumb when checked at the stud edges. And they must be plumb when checked at the stud faces with all studs parallel. The bottom plate may be concrete-nailed to the floor and the top plate to either the bottom edges of the first floor joists or to inter-joist blocking (should the joists run parallel with your proposed wall).

Mark out both top and bottom plates together for accuracy, on 16-inch centers. Now, locate and nail up the top plate. Drop plumb lines from both ends of this plate to locate the bottom plate, but don't nail it down just yet. Instead, fit and toenail the studs to the marked centers of both plates and then adjust the final position of the bottom plate with the aid of a level before nailing.

Mount the electrical box(es) for the outlets and make sure that the vent tube, hoses, and any gas connections will clear all studs when the appliances are brought up to the wall. Staple up the insulation (foil side facing inwards), and nail on the drywall. Finish up the wall by installing the panels with the appropriate color panel nails, cutting out for boxes and tubes as you go. Close up the electricals after inspection.

Suspended ceiling

We found it easiest to install the suspended ceiling framework at this time—no reaching behind cabinets or climbing over appliances. In addition to the 2×4-foot panels, a suspended ceiling requires wall angle, main runners, crosspieces, and suspension wires. Try to leave about a 4-inch space between the joists and the ceiling framework so you'll be able to insert the panels easily.

Begin the ceiling by mounting the wall angle, using box or drywall nails. A stretched string and line level will help you to get a straight, horizontal installation. Continue this procedure around the perimeter of the room. Wall angle (runners and crosspieces as well) can easily be trimmed to length with a metal snips.

In order to center the panels in the room, follow the method described in an accompanying box; then trim the ends of the main runners and locate them the appropriate distances from the walls parallel to them. Now, suspend the main runners with wires every 4 feet of their length and fill in

Before conversion, the laundry center was nothing more than a washer and the dryer dumped near the necessary outlets in the basement. No counters, no storage cabinets, no amenities. Page opposite—the same center after the conversion. Now appliances are flanked by utility cabinets. There's space for sorting and folding clothes and even a rolling hamper.

the structure with 4-foot crosspieces, making a grid of 2×4-foot openings, except as calculated at the borders. Then, adjust the structure with the wires to eliminate any sags or bulges, making sure that the openings are square. Drop in the panels, and you're done with this major job.

Cabinet construction

The appliances in our laundry have a combined width of 55 inches; if your washer and dryer have the same combined width, use the dimensions given on the drawing for the right-side cabinet. If the combined width of your appliances is other than 55 inches, make the calculation given in an accompanying box to determine the actual width of the cabinet; then modify the widths of the top, bottom, back, divider, front, and shelf to accommodate the change.

The cabinet construction uses plywood surfaces and pine trim, assembled with white glue and nails. Cut out all cabinets and assemble them, except for the face frames. Fill and sand all exposed edges, or—even bet-

ter—edge them with pine. Paint the cabinets, then hook up the appliances. Install the lower cabinets next to the appliances, allowing ¼-inch spacing between the appliances and between the appliances and the cabinets. Place the upper cabinet in position and clamp it to the outer sides of both left- and right-hand cabinets with the front edges flush (to receive the face frame). Screw or nail all cabinets to the wall. Screw or nail the upper cabinet to the outer side extensions and, for additional support, install a prepainted cleat under it. Trim the face frame pieces to fit and install them with glue and nails. Set all nails, fill all holes and sand them smooth before touching up the paint.

Hang the prepainted doors covering the end compartments of the upper cabinets with self-closing hinges designed for overlay doors and add the door pulls. Install a pull grip to the front face of the roller cabinet and check the cabinet's alignment with the left-side, lower cabinet. Minor discrepancies can be corrected by shimming between the roller cabinet bottom and the castor mounting pads.

Upper Cabinet Dimensions

TOP SHELF	¾ × 12 × 94½	PLYWOOD
BOTTOM SHELF	¾ × 12 × 94½	PLYWOOD
DIVIDERS	¾ × 12 × 18	PLYWOOD
SIDES	¾ × 12 × 20¼	PLYWOOD
BACK	¼ × 21 × 96	PLYWOOD
NAILERS	¾ × ¾ × 12	PINE
FACE FRAME	¾ × 21 × 96*	PINE
DOORS	¾ (L & W TO SUIT)	PLYWOOD

*FACE-FRAME DIMENSIONS NOT INCLUDING LEFT AND RIGHT SIDE VERTICAL STILE EXTENSIONS TO LOWER CABINETS

Left-Side Cabinet Dimensions

OUTSIDE	¾ × 28 × 76	PLYWOOD
INSIDE	¾ × 28 × 35¼	PLYWOOD
TOP	¾ × 21 × 28	PLYWOOD
BOTTOM	¼ × 19½ × 27¼	PLYWOOD
BACK	¾ × 19½ × 35¼	PLYWOOD
NAILERS	¾ × ¾	PINE

Roller Dimensions

BACK	¾ × 18 × 27	PLYWOOD
FRONT	¾ × 18 × 27	PLYWOOD
SIDES	¾ × 24 × 27	PLYWOOD
FRONT FACE	¾ × 21 × 35¾	PLYWOOD
BOTTOM	¾ × 16½ × 24	PLYWOOD

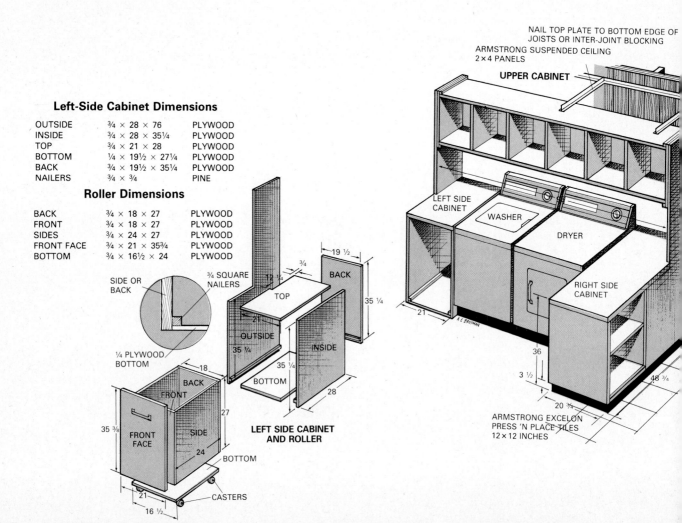

LEFT SIDE CABINET AND ROLLER

The floor

Center the floor tiles using the same method as with the ceiling. Before laying them, make sure the surface is smooth and free of dust. Minor bulges can be hammered down and hollows may be filled with cement patch (no aggregate).

Snap a chalk line on the floor at a convenient location according to your centering calculation, then simply peel off the protective papers and press the tiles in place. Butt them tightly to avoid any gapping. A scissors or shears will handle any trimming to an edge or around obstacles. Apply some compatible colored molding where applicable—*Bernard Price.*

Calculations

Panels to center ceiling. Measure the length of the room in inches. Divide this number by the width of the panel you're using in that direction. You'll get a number and a remainder. The whole number represents full-size panels. Add the remainder to the width of the panel you're using in that direction and divide the sum by two. The result is the width of the border panels. Do the same type of calculation for the adjacent wall. Example: The panel width is 48 inches and the wall dimension is 136 inches. Divide 136 inches by 48 inches and get two full panels, plus 40 inches. Add 40 inches to 48 inches to get 88 inches. Divide by two and get 44 inches for the width of border panels.

Width of the right-side cabinet. Add the combined width of your washer and dryer to ¾-inch spacing requirement. Add this sum to 20¼ inches (left-side cabinet minus the thickness of the outer side). Subtract this from 96 inches (width of upper cabinet) and then add ¾ inch (thickness of right-side outer piece) to the result. Example: Combined width of washer and dryer is 56 inches. Add ¾ inch and get 56¾ inches. Add 20¼ and get 77 inches. Subtract 77 inches from 96 inches and get 19 inches. Add ¾ inch to 19 and get 19¾ inches for the actual width of the right-side cabinet.

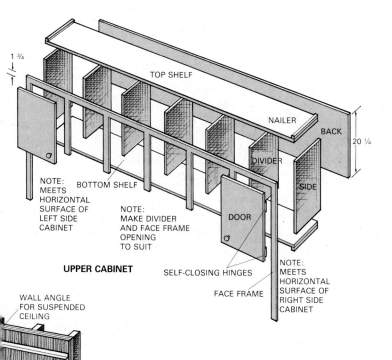

UPPER CABINET

Right-Side Cabinet Dimensions

INSIDE	¾ × 28¾ × 31¾	PLYWOOD
OUTSIDE	¾ × 28¾ × 72½	PLYWOOD
BACK	¾ × 19¼ × 31¾	PLYWOOD
DIVIDER	¾ × 19¼ × 31	PLYWOOD
FRONT	¾ × 20¾ × 31¾	PLYWOOD
SHELF	¾ × 20¾ × 19¼	PLYWOOD
BOTTOM	¾ × 19¼ × 47¼	PLYWOOD
TOP	¾ × 20¾ × 48¾	PLYWOOD
BASE SIDES	¾ × 3½ × 45	PINE
BASE ENDS, CENTER	¾ × 3½ × 11¾	PINE
CLEAT	¾ × 1½ × 15	PINE
FILL	¾ × ¾ × 19¼	PINE

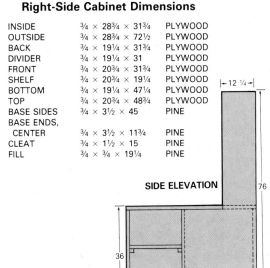

SIDE ELEVATION

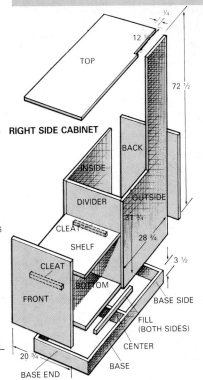

RIGHT SIDE CABINET

utility storage

Utility storage areas eliminate clutter and let you find items fast. Those spare windshield-wiper blades, for example, should be stowed with other car accessories, not buried behind paint cans. Storage centers can also keep dangerous items—poisons, caustic cleaners, power tools—away from curious children and pets.

These two clever storage centers, built for us by the American Plywood Association, are easy to construct, and will help in organizing your shop, basement, or garage.

One storage center, illustrated at right, is a freestanding shelf unit. This design uses the shelving's own weight and load to hold the unit firmly against a wall. Follow the plans provided to cut three 6½-foot-long shelves from a single sheet of plywood.

The second design, shown below, is a childproof storage center on wheels that can double as a mini-workbench. Shallow shelves on one side keep small cans and bottles readily accessible. The perforated-hardboard insert has hooks for paintbrushes or small garden tools. The opposite side of this mobile center has four divided shelves, each almost a foot deep, that provide plenty of room for buckets, air filters, paint trays, and rollers.

Open or closed, the storage center will roll into a corner or push against a wall. The top has more than 4 feet of work surface for quick paint jobs or potting plants. Remove the lock and the unit becomes a toybox. Assembly? Just build the two boxes and hinge them together—*Bob Wilson.*

MATERIALS LIST

Freestanding shelves
A-C exterior or A-D interior APA grade-trademarked plywood—one 4 × 8 × ¾-inch sheet
Standards and brackets—four 8-foot 2 × 4s
Nails and white glue

Rolling chest
A-C APA grade-trademarked plywood—two 4 × 8 × ½-inch sheets
Perforated hardboard—2 × 4 × ½-inches
36-inch continuous hinge, five industrial rubber-wheel casters, padlock hasp and turnknob, nails, and glue

Rolling Chest

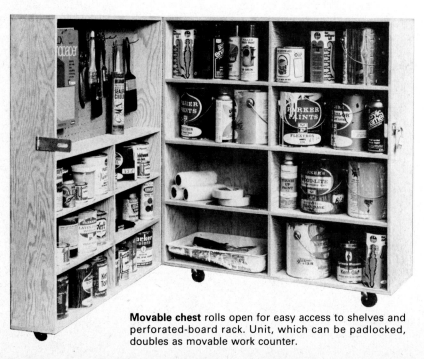

Movable chest rolls open for easy access to shelves and perforated-board rack. Unit, which can be padlocked, doubles as movable work counter.

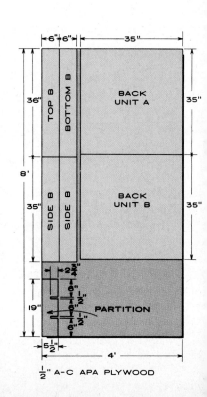

½" A-C APA PLYWOOD

Freestanding Shelves

Top-left isometric diagram labels

- ¾" PLYWOOD SHELVES
- 19½"
- 6'-6"
- 16"
- 3½"
- ¾" PLYWOOD GUSSETS
- 38¼"
- 2 x 4's 5'-11¼" LONG
- 38¼"
- 4's LONG

¾" Plywood Cutting Diagram

- GUSSETS
- 8'
- 6'-6"
- 9"
- 9"
- 3½"
- 9"
- 1½"
- SHELVES
- 9"
- 9"
- 9"
- 9"

¾" PLYWOOD CUTTING DIAGRAM

Instant storage on nearly 20 feet of shelves makes this unit especially handy for a masonry wall. Even when empty, shelves do not need anchoring.

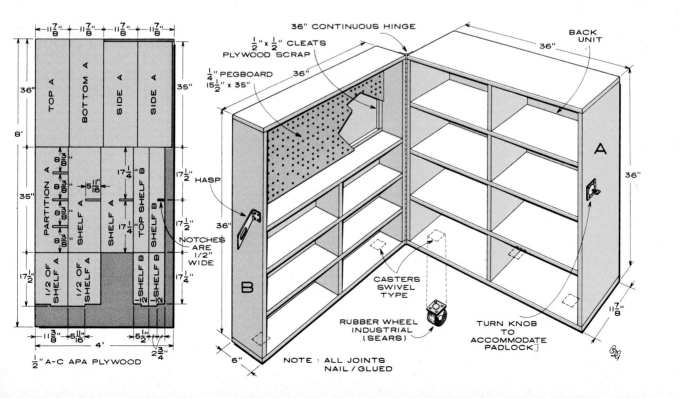

Bottom-left cutting diagram (½" A-C APA plywood)

- 11⅞" 11⅞" 11⅞" 11⅞"
- 36"
- TOP A
- BOTTOM A
- SIDE A
- SIDE A
- 35"
- 8'
- PARTITION A
- ¾" 3"
- 5 11/16"
- SHELF A
- SHELF A
- SHELF A
- TOP SHELF B
- SHELF B
- 17¼"
- 17½"
- 35"
- 17½"
- 17¼"
- 1/2 OF SHELF A
- 1/2 OF SHELF A
- ½ SHELF B
- SHELF B
- ½ SHELF B
- 17½"
- 17¼"
- NOTCHES ARE ½" WIDE
- ¾"
- 11⅞"
- 5 11/16"
- 4'
- 5 11/16"
- 2¾"
- ½" A-C APA PLYWOOD

Bottom-right isometric cabinet diagram

- 36" CONTINUOUS HINGE
- BACK UNIT
- 36"
- ½" x 1" CLEATS PLYWOOD SCRAP
- ¼" PEGBOARD 15½" x 35"
- 36"
- 36"
- A
- 36"
- HASP
- 36"
- B
- CASTERS SWIVEL TYPE
- RUBBER WHEEL INDUSTRIAL (SEARS)
- TURN KNOB TO ACCOMMODATE PADLOCK
- 11⅞"
- 6"
- NOTE : ALL JOINTS NAIL / GLUED

table-saw master jig

Ever wish you could make your table saw do more? With a master jig, you can. The multipurpose jig shown here turns your saw into a versatile yet precise tool. In fact, once you use it for a while, you'll wonder how you managed without it.

Its secret? Since the work and jig table move together, sawing is easier and more accurate. And attachments for mitering, feathering, splining, tenoning, slotting, and many other standard—and not so standard—operations make it one of the most useful accessories going.

The jig is shown on a Rockwell 10-inch Unisaw, which has a 27×36-inch table. Many other 9- and 10-inch machines (the most popular sizes) are similar, so the dimensions in the drawings are generally applicable. The only tailoring you need to do is on the dimensions of the bars and their placement on the main table. The bars slide in the table slots, and their positions vary from saw to saw.

The thickness of the sliding table reduces the maximum blade projection, but since projection is normally $2\frac{3}{4}$ to $3\frac{1}{2}$ inches, reducing it by $\frac{1}{2}$ inch isn't critical; it'll still cut most standard stock.

Accurate construction of the jig is important, although some tolerances are built in. For example, the fastening holes in the attachments are $\frac{5}{16}$ inch in diameter, even though they are secured with $\frac{1}{4}$-inch locking hardware threaded into $\frac{1}{4}$-inch T nuts. This permits $\frac{1}{16}$-inch adjustments.

Use a good grade of maple or birch plywood. After each part has been sized and the corners rounded, use fine sandpaper to smooth surfaces and edges. For the table top, cut a sheet of aluminum to size and bond it to the table using contact cement.

Next, shape the bars to fit your table slots. Sand the bars so that they slide easily in the table slots without wobble. Put the bars in position, then place the sliding table so it's centered over the saw blade. Make sure its edges are parallel to the table slots. Use C-clamps to hold the bars to the sliding table, and drive a short brad to keep them in place. Repeat this procedure at the opposite ends of the bars, and permanently attach them with $\frac{3}{4}$-inch #8 flathead wood screws. Be sure to drill shank holes for the screws; if you don't, driving the screws may spread the bars and cause them to fit too tightly in the table slots.

Next, form the saw kerf. Work with a good saw blade,

Table-saw master jig consists of an aluminum-veneered sliding table with attachments for a variety of shop jobs. Shown crosscut fence, adjustable stop, vertical work support, vertical miter guides, hold-down, and right-angle guide (mounted on table).

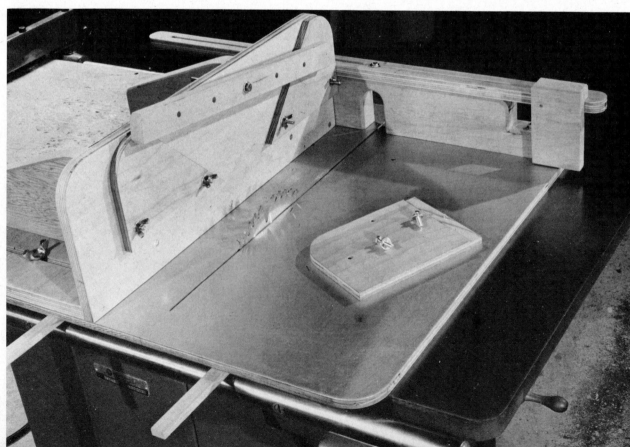

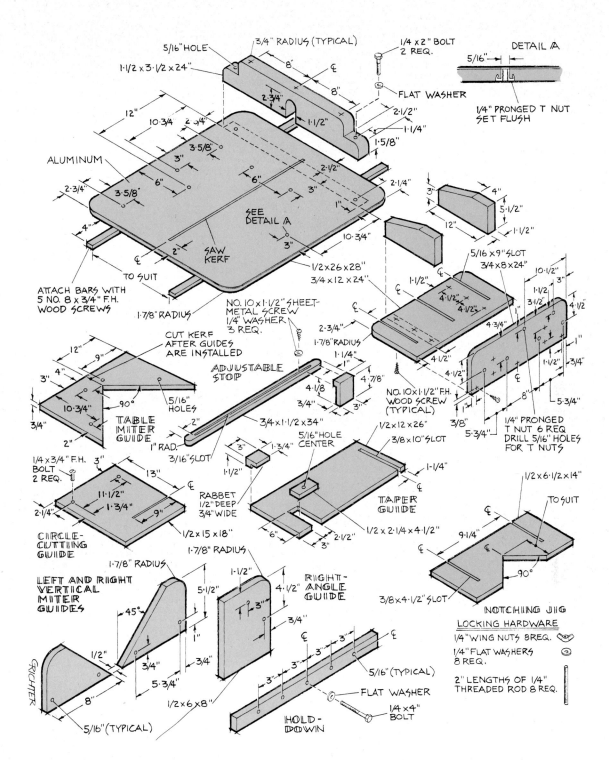

one that you will always use with the master jig. Don't use a conventional hollow-ground blade—the best kind to use is a carbide-tipped combination blade.

With the machine off, set the blade to its lowest point and put the sliding table in place. Turn on the machine and slowly raise the blade until it cuts through the sliding table, then move the table to lengthen the kerf.

Carefully mark the locations for all the T nuts needed for the table. Work with a scriber, but mark lightly so you don't mar the aluminum. Mark the hole locations with a prick punch and then drill a ⅟₁₆-inch hole at each mark. Use a ¾-inch brad-point bit on the underside to form a ⅟₁₆-

inch-deep counterbore and then, from the top side, open up each of the holes to ⁵⁄₁₆ inch.

Install the ¼-inch T nuts by tapping them into place with a hammer. They must be flush with, or slightly below, the surface of the plywood.

Shape the crosscut fence, and then drill the ⁵⁄₁₆-inch holes for the bolts used to secure the fence to the table. The three hole locations on top of the fence are for the screws that hold the adjustable stop when it's in use.

Rather than being a one-piece V-block, the miter guides are made in two pieces. You can use both pieces when mitering parts that have been precut to length or just one

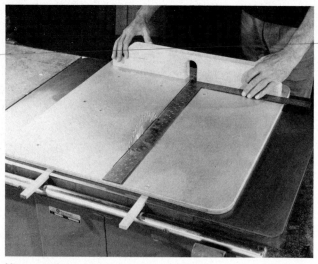

Use a square to set crosscut fence 90 degrees to the blade before tightening fence bolts. Difference between ¼-inch locking bolt and ⁵⁄₁₆-inch holes in fence allows for minor adjustment.

To make consecutive cuts along a single length of stock, remove the crosscut fence and work with one guide. If stock can't be turned over, you can alternate left- and right-hand guides.

Use the adjustable stop to saw multiple pieces to the same length. This attachment permits cuts up to 28 inches long. Hold work firmly and return to starting position before removing.

Use table miter guides to cut slots in rounds. To mark center of round or square stock, lower blade to minimum and make two cuts at right angles; center is where the kerfs intersect.

With table miter guides in place you can form accurate miters on precut stock—even molding, since you can cut on either side of blade. Check position with triangle before locking.

Vertical right-angle guide is used for end cuts such as slots and tenons. Before securing in position, check with a square to be sure guide's bearing edge is 90 degrees to the table.

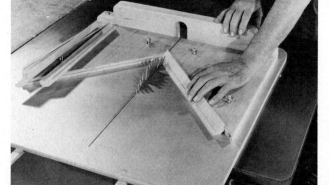

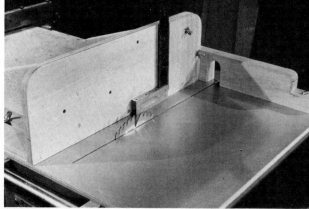

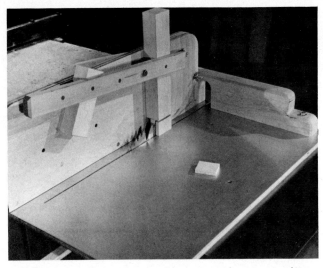

To form tenon, first make shoulder cuts and set up as shown for cheek cuts. For second cuts on each you just reverse stock's position. Use same-size scrap under free end of hold-down.

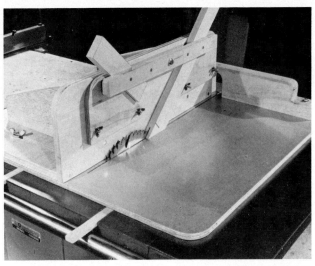

For centered spline groove, reverse stock for second cut. For off-center groove, use both guides. Face matching surfaces toward guide, cut both pieces, reset guide, and pass again.

To form a slot in narrow work, set the vertical work support to the width you want the shoulder of the slot to be. Make a pass, reverse stock, and make another—the slot will be centered.

To set vertical miter guides, first lock one into place, checking with a triangle to be sure bearing edge is 45 degrees. Secure the other guide after checking with square as shown.

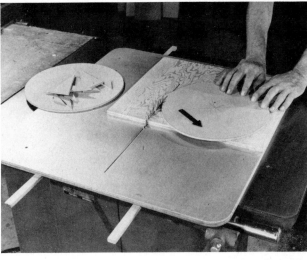

To cut circles, nail stock to guide and make series of straight cuts, rotating work 10 degrees with each pass. Then align nail with blade front and rotate work in direction shown.

Notching jig can be used to make odd-shaped pieces that would be difficult to form with standard tools. Shape of jig varies according to what you need to make with it.

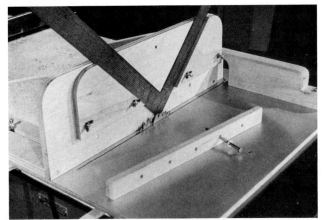

SPECIAL GUARD

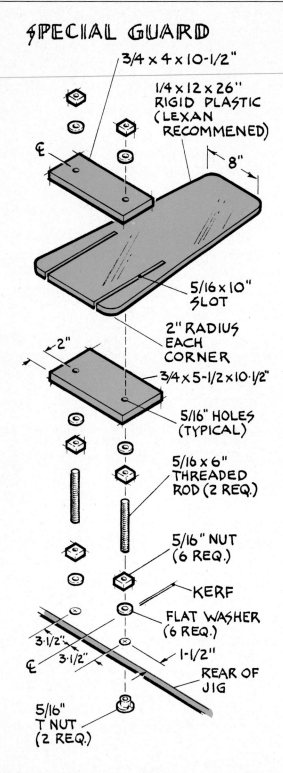

3/4 x 4 x 10-1/2"

1/4 x 12 x 26" RIGID PLASTIC (LEXAN RECOMMENED)

8"

5/16 x 10" SLOT

2" RADIUS EACH CORNER

3/4 x 5-1/2 x 10-1/2"

2"

5/16" HOLES (TYPICAL)

5/16 x 6" THREADED ROD (2 REQ.)

5/16" NUT (6 REQ.)

KERF

FLAT WASHER (6 REQ.)

1-1/2"

REAR OF JIG

3-1/2"

3-1/2"

5/16" T NUT (2 REQ.)

A special guard. The only required modification of the jig is the addition of the two holes for the ⁵⁄₁₆-inch Tee-Nuts. The posts (⁵⁄₁₆-inch threaded rod) thread into the Tee-Nuts and are secured with a washer and nut. The guard parts mate as shown in the drawing.

The guard is adjustable vertically and longitudinally so the blade and the cut being made will always be covered by the plastic shield. Adjust the height of the guard so it is between ¼ inch and ½ inch above the blade. The width of the shield keeps your hands away from the cutting area.

The guard is used for all operations done with the sliding table. It is not used with the vertical work support. But with this accessory, hands don't have to be placed in a hazardous position.

when cutting to size. You can remove the crosscut fence and use either the left- or right-hand guide when making consecutive miter cuts on a single length of stock.

Mark the 45-degree angle on each guide by using a combination square. Saw approximately to the mark and finish by sanding exactly to it. Lock the guides on the table so that the short edges abut and the joint is centered over the kerf. Then move the table so the saw blade spreads them apart. When the guides are mounted, use a triangle to be sure they are in correct alignment with the saw kerf.

To make the vertical work support, first make the base—the slots can be formed by making repeated passes with the saw blade—and then add the braces. Check with a square to be sure the front edges of the braces are 90 degrees to the base.

Shape the face and carefully lay out locations for the T nuts. Here, the T nuts do not have to be set flush. All you need to do is drill a ¹⁄₁₆-inch pilot hole and then enlarge it to ⁵⁄₁₆ inch.

Mark the 45-degree angle on the vertical miter guides by using a combination square as shown in the illustration.

To form the long slot in the adjustable stop, first drill ³⁄₁₆-inch end holes and then cut between them with a coping saw or jigsaw. The slots in the other attachments can be formed by repositioning the parts and repeating passes on the table saw.

Sand all parts before assembly. Apply two to three coats of sealer, sanding between coats and after the final one. Apply paste wax to the saw table and to the bars and underside of the sliding table, then rub to a high polish— *R. J. Cristoforo.*

Materials list
Sliding table
Main table: One ½ × 26 × 28-inch cabinet-grade plywood
Table cover: One 26 × 28-inch aluminum sheet
Bars: Two ⅜ × ¾ × 36-inch hardwood
Crosscut fence: One 1½ × 3½ × 24-inch hardwood; nine ¼-inch pronged T nuts; ten ¾-inch #8 flathead wood screws; two ¼ × 2-inch bolts; two ¼-inch flat washers
Vertical work support
Base: One ¾ × 12 × 24-inch cabinet-grade plywood
Face: One ¾ × 8 × 24-inch cabinet-grade plywood
Brace: Two 1½ × 5½ × 12-inch hardwood; six ¼-inch pronged T nuts; thirteen 1½-inch #10 flathead wood screws
Table miter guides
Guide: Two ¾ × 12 × 16-inch cabinet-grade plywood
Adjustable stop
Bar: One ¾ × 1½ × 34-inch cabinet-grade plywood or hardwood
Stop: One 1¼ × 3 × 4⅞-inch hardwood; three 1½-inch #10 sheet-metal screws; three ¼-inch flat washers
Circle-cutting guide
Platform: One ½ × 15 × 18-inch cabinet-grade plywood; two ¼ × ¾-inch flathead bolts
Taper guide
Platform: One ½ × 12 × 26-inch cabinet-grade plywood
Stop: One 1½ × 1¾ × 3-inch hardwood
Clamp pad: One ½ × 2¼ × 4½-inch hardwood or plywood
Vertical miter guides
Guides: Two ½ × 5½ × 8-inch cabinet-grade plywood
Right-angle guide
Guide: One ½ × 6 × 8-inch cabinet-grade plywood
Hold down
Bar: One ⅝ × 1½ × 18-inch hardwood; one ¼ × 4-inch bolt; one ¼-inch flat washer
Example notching jig
Guide: One ½ × 6½ × 14-inch cabinet-grade plywood
Locking hardware
General-use pieces: Eight ¼-inch wing nuts; eight ¼-inch flat washers; eight ¼ × 2-inch lengths of threaded rod

jigs for belt and disc sanders

Accessories make the tool. That applies to the combo belt/disc sander just as it does to any other machine. Jigs can increase the tool's capabilities, make it easier to use, improve accuracy, and ensure duplication when you do a production run.

Sometimes accessories are available at extra cost. But many times the jigs you need just *aren't* available, so you must design and construct them. That needn't be a hardship; in fact, it can be fun. The jigs and techniques shown here are just as usable if your belt and disc sanders are separate machines.

Check the dimensions on the drawings against your tool; some minor changes may be in order: Your table may be smaller, it may not have a miter-gauge slot, your belt sander may

have a stop instead of a table. In the latter two cases, you can exclude the miter-gauge bar I put on some jigs and just clamp the jig's platform in place. Most belt sanders have attachment holes for commercial accessories; use these to attach your homegrown versions. If your machine has no such holes, just clamp on the jig instead. Be sure your jigs are elevated above the abrasive belt: Use thin wooden shims or heavy washers between the tool and the jig.

One general thought: A sander should be regarded as a finishing tool. Except in special cases, it should not be used to *shape*. Always make your saw cut as close to the line as possible. Leave sanding for the final touch—*R. J. Cristoforo.*

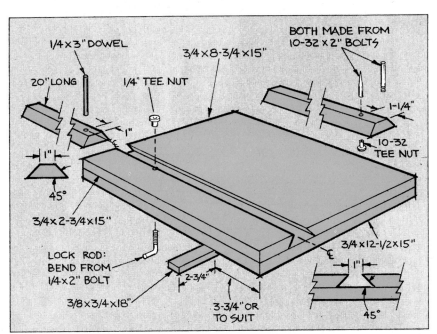

Circle-sanding jig for a disc sander is a pivot point mounted on a sliding bar that fits a 45-degree dovetail cut on the inside edges of the platform's top pieces. The distance from pivot to disc (the radius of the circle) will be constant. You mount the work and move the bar until work touches disc. Then you lock the bar in place (see drawing) and rotate the work to sand its edge. When you need duplicates (as for wagon wheels), mark the bar and platform to reposition the bar accurately. The dowel pivot (photo) requires a center hole in the work; a pointed pivot (drawing detail upper right), made by chucking a headless bolt in a drill and filing it, lets you sand work without a hole. The slide bar is reversible so you can have both pivots—one on each end.

Sand straight pieces to width by passing the work between a clamped-on fence and the "down" side of the disc. The fence is offset enough to let the work contact as much of the "down" side as possible without touching the "up" side. Arrow shows the feed direction; the disc turns counterclockwise. Start by placing the work on the table, snug against the guide. Then slowly move the entire work past the disc. For wide work, clamp the guide to an auxiliary table secured to the sander's table.

Chamfering jig for dowels is a length of 2×4 drilled for various dowel diameters. Clamp the jig to the sander's table at the angle you need, pass the dowel through the appropriate hole, then rotate it slowly against the abrasive. A light touch is best. You can form bevels or flat-sided points by holding the dowel steady as you move it forward. Size the holes in the jig accurately—the dowels should move easily but without wobble. Try coating tight-fitting dowels lightly with paste wax.

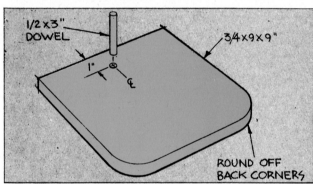

Sand curves to width with a guide made of a dowel installed in a platform that you clamp to the sander's table. Distance from the inside edge of the dowel to the disc equals desired width of the work. The work must be constantly turned to maintain the contact point. First you must carefully sand the *inside* edge of the curve—on a drum sander or on the belt sander's top drum.

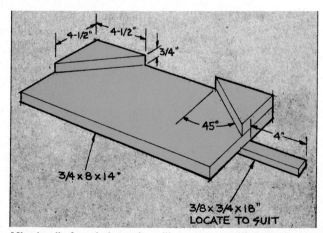

Mitering jig for a belt sander will assure you a lifetime of accurate 45-degree miter cuts if you make it carefully. Important factors: (1) Be sure the sander's table is 90 degrees to the belt. (2) Locate the jig's platform so its forward edge is parallel to and about ⅛ inch from the belt. (3) Position the guide blocks by working from the belt, not from the edge of the platform. The stock can be sanded flat or on edge. There's both a left- and right-hand guide block, so you can sand stock that can't be flipped.

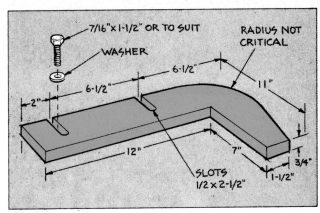

L-shaped stop lets you use most of the sanding area when the belt is horizontal. With the regular table or stop you can use only a section of it. The jig is attached with bolts put through existing holes, or with clamps. This is a good jig to use when sanding pieces that are no longer than the platen (the metal plate that backs up the belt). Hold the work firmly against the stop. Little pressure is needed; the work's weight is usually enough. Be careful of your hands, especially on thin pieces.

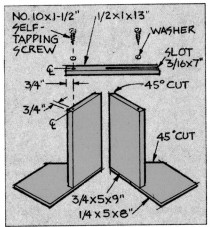

Adjustable chamfering jig lets you control the angle and amount of chamfer. Place the work on the sander's table and against one of the jig's uprights. Then move it forward to contact the abrasive. The opposite upright acts as a stop to control how much material is removed. When setting up this jig, use the top brace to lock the angle of the uprights and the opening between them. Then attach the jig to the sander's table with two C-clamps. Work can be held flat or on edge so it's easy to uniformly chamfer all four edges of a piece. Slight chamfers can be done by sanding only; heavy chamfers should be formed first by sawing. When you have a lot of work to do, shift the jig occasionally so you don't work on one slim section of the belt.

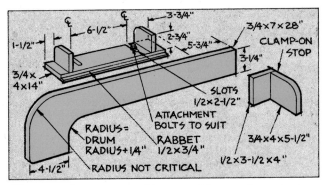

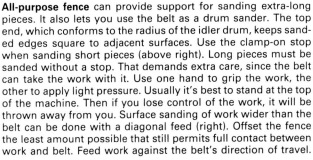

All-purpose fence can provide support for sanding extra-long pieces. It also lets you use the belt as a drum sander. The top end, which conforms to the radius of the idler drum, keeps sanded edges square to adjacent surfaces. Use the clamp-on stop when sanding short pieces (above right). Long pieces must be sanded without a stop. That demands extra care, since the belt can take the work with it. Use one hand to grip the work, the other to apply light pressure. Usually it's best to stand at the top of the machine. Then if you lose control of the work, it will be thrown away from you. Surface sanding of work wider than the belt can be done with a diagonal feed (right). Offset the fence the least amount possible that still permits full contact between work and belt. Feed work against the belt's direction of travel.

backyard solar-heated workshop

It's no fun to work in a backyard shop in the middle of winter. That's what I realized after I designed and built my first workshop. Though I'd designed that first shop only for warm-weather use, once I had it I found that I wanted to use it throughout the year. During winters in upstate New York, where I live, keeping the shop barely warm required constant feeding of a wood stove. When I recently moved a few miles from my former residence, I determined to build a shop that would be convenient to work in all year long.

The energy crunch and my interest in small, efficient buildings led me to investigate solar heating, even though my site was less than ideal. After studying the major books in the solar-heating field, I decided that the best solution for me was a direct-gain passive system. This arrangement makes the most sense for a small, inexpensive structure. Since the building itself is the collector, there's little added cost over normal construction materials, no extra for solar hardware.

In my design, sunshine passes through large fixed windows and heats a dark concrete floor slab and the earth below it. To retain the heat, the floor slab and foundation are insulated from the surrounding earth around the perimeter. Ceiling and walls have only normal insulation, and doors and shutters are weatherstripped as usual. The only moving parts of my heating system are the folding, insulated shutters that cover the windows.

At my site, on short winter days the sun is full on the workshop at 9:30 in the morning. Some shade trees (I am as yet unwilling to cut them down) filter most of the afternoon sun after 1:30, so the shop misses about two hours of midwinter sun. I calculated that due to my site restrictions I am garnering only 60 to 65 percent of the available winter insolation. Even so, I think my results are impressive.

Solar performance

After I open the shutters each morning, the workshop warms up quickly. The inside temperature will rise as much as 30 to 35 degrees by 1:30, and I close the shutters to hold the heat. The gain is 7 to 10 degrees per hour on a bright day. This gain depends on the clarity of the sky, not the outside temperature (see table). After the shutters are closed, the temperature in the shop falls an average of 1 to 2 degrees per hour overnight.

This means that the shop is not only comfortable to work in until late in the evenings, but the heat gain is cumulative. After several successive sunny days, when I open the shop to the morning it will be 10 to 20 degrees above the outside temperature. The shop reaches a top temperature of about 65 degrees if it's 25 outside. When it's 40 outside and sunny, the shop temperature rises well into the 70s. I calculate that, with optimal siting, inside temperatures of at least 70 degrees could be achieved in 25-degree weather.

As a backup when the temperature falls below zero or the sky stays cloudy for several days, I installed a small wood stove, an airtight Petit Godin No. 3731. I found an outside air-supply vent necessary, though, to keep the stove from gulping heated shop air.

Solar planning

For the solar design, I used a rough rule of thumb of 1 cubic foot of storage material to 1 square foot of glass collector. After seeking some expert advice, I reduced the glass area to avoid overheating in midday sun when outside temperatures wouldn't be too low—say, in the 40s. I had planned to pour a 1-foot-thick concrete slab, but, after figuring the total expense of that much concrete at today's prices, I reconsidered. I remembered that some solar projects have included earth as a heat sink under a floor and found it to be almost as good as masonry in holding heat. I decided to make a thinner slab and depend on the gravel and earth underneath to store any extra heat.

The shed shape was chosen for sim-

Shed framing can be a one-man job. Workshop's walls were assembled on slab, raised, and braced by author alone. Short 2×8 rafters required two cutoffs.

Aluminum paint on inside of shutters reflects extra light into shop. Shutters hang from jamb on 4-inch strap hinges, with 2-inch butt hinges between panels.

Interior of shop (above) has space for Sears table and band saws and plenty of counter area. Concrete floor is painted flat black to absorb sunlight entering window. Workshop actually heats up best when snow reflects extra sunshine. Two doors to shop are under covered porch: small entry door to minimize heat loss, and extra-wide one for bulky materials and summer ventilation.

plicity of construction and function. The sloped ceiling would move warm summer air up to the belvedere where it could be vented. The high south side also makes the shop seem much bigger and more comfortable.

Construction

To lay the foundation, I had a backhoe dig a 1-foot-wide, 3½-foot-deep trench around the shop and porch area. A 2-foot-wide trench of the same depth was then cut to separate the porch and shop slabs. Two-by-fours were set around the outside of the trench and down the center of the 2-foot trench and leveled several inches above grade. Tongue-and-groove, 2-inch-thick blue Styrofoam—the only type specifically for insulating below grade—was taped to the inside of the 2 × 4s flush to the top, creating a form for the concrete. Forms for the piers were made from plywood, but Sonotube, a ready-made cardboard product available through most lumberyards, would work as well. After the pier forms were in place, the ditch was partly backfilled and tamped, and gravel was laid over the fill (see drawing). Steel reinforcing bar in the piers and edge beams was then wired in position and 6 × 6 wire mesh laid to strengthen the slab floors.

The foundation and slabs were filled in a single pour to the top of the Styrofoam. Half-inch bolts for anchoring the 2 × 4 sills were set in the slabs near the edge. A pressure-treated 2 × 4 was used in the form between the porch and shop area and left in place after pouring.

Walls of 2 × 4s on 24-inch centers were assembled on the slab and raised into place (except for the south wall, which was built in place after the end walls were up). Double 2 × 4s were used around the door and fixed windows. However, built-up 2 × 3 posts were used between shutter jambs, so the insulated shutters could fit flush to the outside wall when closed. The workshop and doors were faced with ⅝-inch T1-11 plywood (grooved 4-

Foundation Styrofoam extends at least 3 feet into ground. Upper 12 inches of insulation is covered on outside with fiberglass sheet for protection (see detail). Foundation plans of piers and edge beams can't be used in poor bearing soil; this condition requires wider footings along the perimeter below freezing depth. Air supply for stove is 1½-inch plastic pipe. Stove's metal flue should be insulated most of its length and a damper added. Electricity is supplied from house via underground rigid plastic conduit.

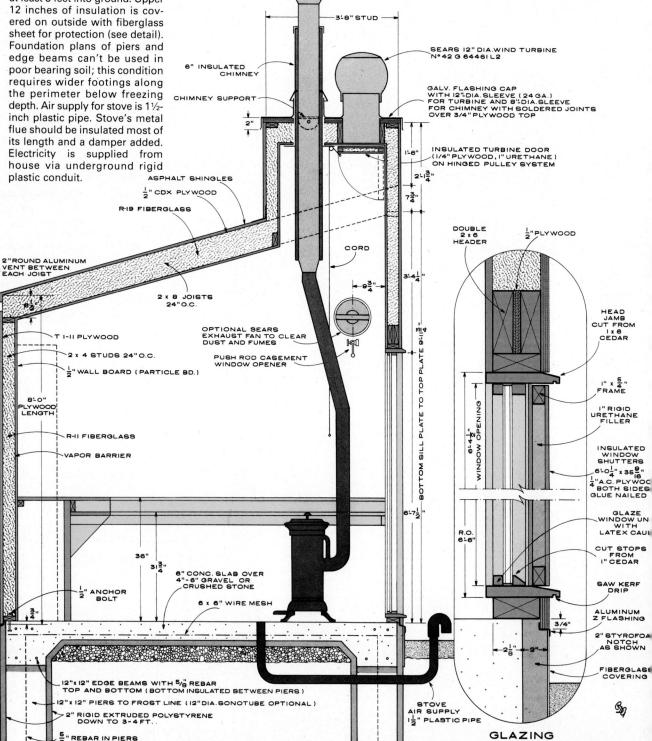

SECTION

GLAZING
AND
SHUTTER DETAILS

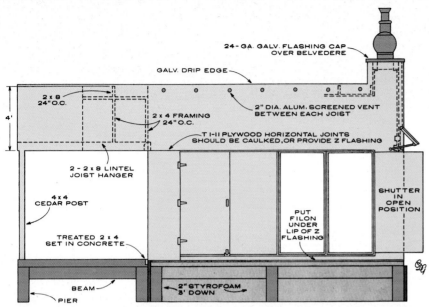

Two-inch-diameter air vents high in wall between rafters keep insulation moisture-free. Glass in fixed windows is tempered, double-glazed, sliding-door replacement. These replacements are the cheapest source of large double-glazed stock.

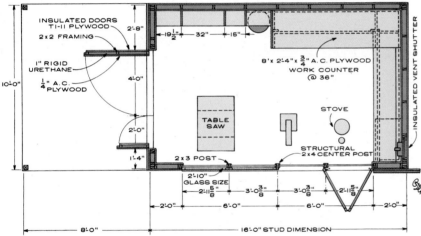

Porch doors and folding shutters are insulated with rigid urethane. Doors are framed with 2 × 3s and covered with siding outside, ¼-inch AC plywood inside. Shutter framing is ⅝-inch stock with ¼-inch plywood glued and nailed on both sides.

WORKSHOP'S PERFORMANCE ON SUCCESSIVE MIDWINTER DAYS

	Time of day	Outside temperature	Inside temperature	Total gain
Day 1 (partly cloudy)	9:30 AM	21	26	
	1:30 PM	21	56	30
	7:00 PM	15	40	
Day 2 (sunny)	9:30 AM	17	30	
	1:30 PM	26	62	32
	7:00 PM	25	47	
Day 3 (hazy)	9:30 AM	27	34	
	1:30 PM	38	55	21
	7:00 PM	38	46	
Day 4 (sunny)	9:30 AM	32	38	
	1:30 PM	44	61	23
	7:00 PM	32	52	

Workshop gains 7 to 10 degrees per hour on sunny days, 9:30 to 1:30, when site limits insolation. Gain is cumulative on successive days, for higher peak temperatures.

inches off center) nailed directly to the studs. As a siding material, T1-11 dispenses with the need for a separate sheathing. Although foam sheathing is being used to increase insulation in some construction, I didn't think it necessary for my small structure.

The shed roof was framed with 2 × 8 rafters. Insulation between rafters was fiberglass R-19, with R-11 in the walls. A heavy plastic vapor barrier was stapled over walls and ceiling. Roofing materials were the usual asphalt shingles and felt over ½-inch exterior plywood sheathing. Inside walls were covered with ½-inch particleboard.

Shutters were constructed of ⅝-inch #1 pine stock with 1-inch rigid urethane sandwiched between ¼-inch AC plywood on both sides. Windows are sealed, double-glazed, tempered glass—standard replacement panels for sliding glass doors. The glass panels were fit in their jambs and held with 1-inch stop on either side and caulked around the edges to prevent infiltration.

The belvedere required heading off a single 2 × 8 rafter and framing in the rest of the box structure with 2 × 4s. A Sears wind-driven turbine ventilator was installed at the top of the belvedere for summer ventilation, and an insulated chimney support added beside it for the stove flue.

The roof of the belvedere is a single sheet of galvanized metal with short collars soldered on around holes for the flue and turbine ventilator. The flashing sleeves for each were then slipped over the collars and set in a generous layer of butyl caulk.

An exhaust fan was added on the side wall just below the belvedere to clear solvent fumes and dust. Urethane-insulated covers similar in construction to the shutters were built for the turbine throat and the exhaust-fan opening during winter. To preserve the siding's natural look, I applied a clear water-repellent sealer. The exterior face of the shutters was primed and painted with bright latex house paint—*Jeff Milstein. Drawings by Carl de Groote*

MANUFACTURERS LIST

For more information about products mentioned in this article, write to the manufacturers listed below. **American Plywood Assn.,** Box 2277, Tacoma, WA 98401 (T1-11 siding); **Bow & Arrow Imports,** 14 Arrow Street, Cambridge, MA 02138 (Petit Godin stove); **Dow Chemical,** 2020 Dow Center, Midland, MI 48640 (Styrofoam); **Sears,** Sears Tower, Chicago, IL 60684 (shop tools, fan, and turbine).

cedar-sided shed

The long, slender yard shed in these color photos is both visually and physically a cluster of three enclosures that have separate functions:
● Facing front, an open-topped pen with hinged doors for trash cans. They're out of sight, and secure from neighborhood animals.
● Sandwiched in the middle, an open-sided section that holds more than a cord of firewood. It's built to hold heavy loads (see photo) and roofed to keep wood dry.
● At rear, there's enclosed storage for bicycles, lawn mower, and other garden tools and equipment. It's lockable for security.

This diverse assemblage is particularly striking because it's sided in western red cedar, nailed up at 45-degree angles. Cedar shingles roof the covered sections.

Stapling chicken wire to the inside edges of the trash-can enclosure's framing lets this section double as a compost bin. The chicken wire keeps the compost away from the sides of the enclosure and allows air to circulate. It you opt for this use, you'd be wise to select pressure-treated wood for the framing. In any case, it's advisable to use preservative-treated wood for ground-contact outdoor projects such as this yard shed.

The overall dimensions of the shed were dictated by site requirements; local zoning forbade any structure within 5 feet of the property line, and the clients wanted it placed in a narrow side yard so the shed could double as a privacy fence. You could alter the overall size or add or delete sections to fit your own needs. For example, if you need both storage for trash cans and a compost bin, add a second pen at the rear of the structure, next to the shed. Its doors could face toward the rear or open to the side—*Richard Stepler. Design by Julia Lundy Sturdevant. Photos by Western Wood Products Association.*

Framing for wood storage and shed is complete. Note that units are separated by exterior-grade plywood panels; trash-can enclosure will be attached at left.

Upper shelf, for split wood and kindling, is built of closely spaced 2 × 4s, wide side up. Heavy logs go on the lower shelf—here the 2 × 4s are narrow side up.

Alternating panels of western red cedar applied at a 45-degree angle add visual interest to long, skinny yard shed (bottom). On the house side, there's plenty of clearance for access to wood storage (left) and the enclosed shed (below).

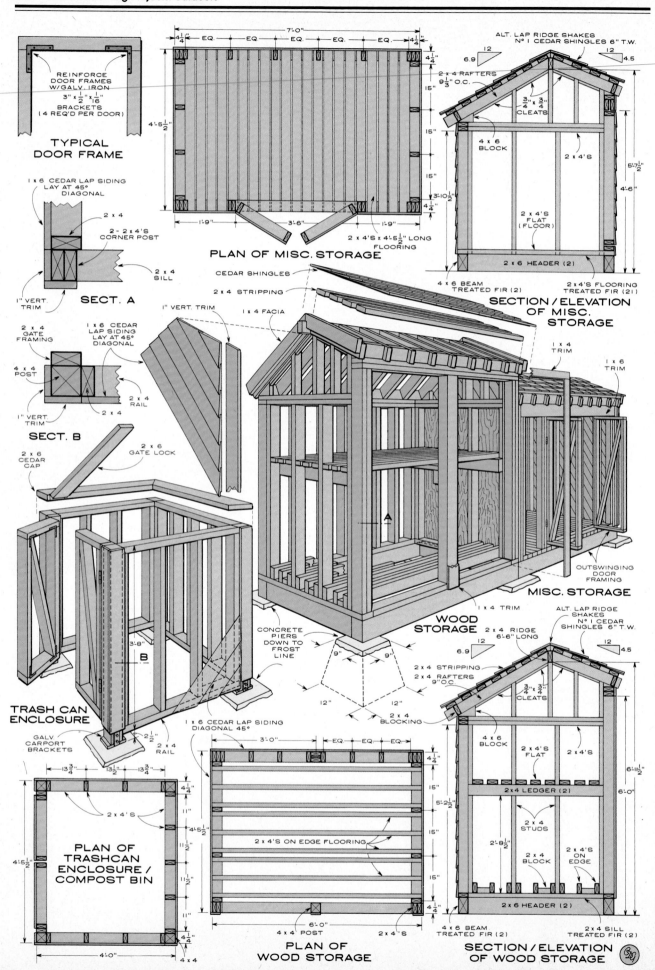

TYPICAL DOOR FRAME

REINFORCE DOOR FRAMES W/GALV. IRON 3" x ½" x 1/16 BRACKETS (4 REQ'D PER DOOR)

1 x 6 CEDAR LAP SIDING LAY AT 45° DIAGONAL

2 x 4

2 - 2 x 4'S CORNER POST

2 x 4 SILL

1" VERT. TRIM

SECT. A

2 x 4 GATE FRAMING

1 x 6 CEDAR LAP SIDING LAY AT 45° DIAGONAL

4 x 4 POST

2 x 4 RAIL

2 x 4

1" VERT. TRIM

SECT. B

2 x 6 CEDAR CAP

2 x 6 GATE LOCK

TRASH CAN ENCLOSURE

GALV. CARPORT BRACKETS

2 x 4 RAIL

2½"

3'-8"

B

PLAN OF TRASHCAN ENCLOSURE / COMPOST BIN

13¾"   13½"   13¾"

4¼"

2 x 4'S

4¼"

11"

4'-5½"

4'-5½"

11½"

11½"

4¼"

4'-0"

4 x 4

PLAN OF MISC. STORAGE

4¼"   EQ.   EQ   EQ.   EQ.   EQ.   4¼"

7'-0"

4¼"

15"

15"

15"

4¼"

4'-5½"

1'-9"   3'-6"   1'-9"

3'-10½"

2 x 4'S x 4'-5½" LONG FLOORING

1 x 6 CEDAR LAP SIDING DIAGONAL 45°

3'-0"   EQ   EQ   EQ

4¼"

15"

5'-2½"

15"

4¼"

2 x 4'S ON EDGE FLOORING

6'-0"

4 x 4 POST   2 x 4'S

PLAN OF WOOD STORAGE

CEDAR SHINGLES

2 x 4 STRIPPING

1" VERT. TRIM

1 x 4 FACIA

1 x 4 TRIM

1 x 6 TRIM

1 x 4 TRIM

MISC. STORAGE

OUTSWINGING DOOR FRAMING

WOOD STORAGE

CONCRETE PIERS DOWN TO FROST LINE

9"   9"

12"   12"

2 x 4 BLOCKING

ALT. LAP RIDGE SHAKES Nº 1 CEDAR SHINGLES 6" T.W.

6.9   12   12   4.5

2 x 4 RAFTERS 9½" O.C.

¾" x ¾" CLEATS

4 x 6 BLOCK

2 x 4'S

2 x 4'S FLAT (FLOOR)

5'-7½"

4'-6"

2 x 6 HEADER (2)

4 x 6 BEAM TREATED FIR (2)

2 x 4'S FLOORING TREATED FIR (21)

SECTION / ELEVATION OF MISC. STORAGE

2 x 4 RIDGE 6'-6" LONG

ALT. LAP RIDGE SHAKES Nº 1 CEDAR SHINGLES 6" T.W.

6.9   12   12   4.5

2 x 4 STRIPPING

2 x 4 RAFTERS 9" O.C.

¾" x ¾" CLEATS

4 x 6 BLOCK

2 x 4'S FLAT

2 x 4'S

2 x 4 LEDGER (2)

2 x 4 STUDS

2 x 4 BLOCK

2 x 4'S ON EDGE

2 x 6 HEADER (2)

6'-11½"

6'-0"

2'-8½"

4 x 6 BEAM TREATED FIR (2)

2 x 4 SILL TREATED FIR (2)

SECTION / ELEVATION OF WOOD STORAGE

how to deck out an above-ground pool

When we decided to install an above-ground pool in our yard, we were determined to make it look as much like an in-ground pool as possible. With the help of Kirk Olson, a twenty-two-year-old telephone repairman with no formal design training, we were able to do all the planning and construction—without calling in the pros.

To minimize the raised look, we established the first level of our deck flush with the bottom of sliding doors from the kitchen. The second, or pool level, would be 7 inches below this. With the levels determined, we could get started.

Artfully designed two-level deck gives above-ground pool the look of a built-in. Top deck level is flush with house entry, making it convenient for outdoor dining and entertaining. Main deck area is 7 inches lower, so swimmers can step from deck into pool. Board-on-board fence screens pool from the street. Safety railing is of simple construction.

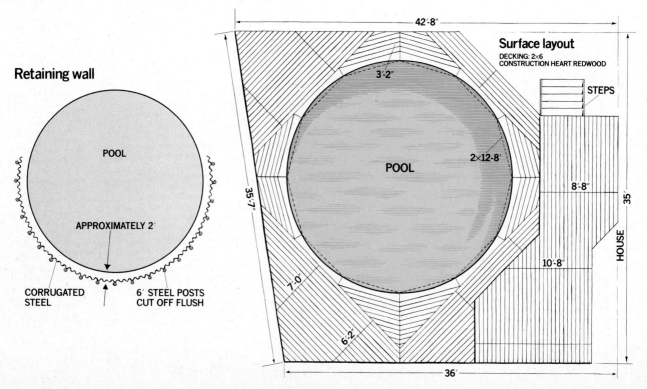

Post and joist layout

42'-8"

4×4 POST

DOUBLE

4×4 POST

DOWN

DOUBLE

LAG TIES

DOUBLE

HOUSE

14'-10½"

14'-9"

35'-7"

FILTER EXIT

DOUBLE

DOUBLE

DOUBLE

DOUBLE

36'

Retaining wall

POOL

APPROXIMATELY 2'

CORRUGATED STEEL

6' STEEL POSTS CUT OFF FLUSH

Surface layout

DECKING: 2×6
CONSTRUCTION HEART REDWOOD

42'-8"

3'-2"

STEPS

POOL

2×12-8'

8'-8"

35'-7"

35'

HOUSE

7'-0"

6'-2"

10'-8"

36'

Doors that provide access to all sides of the pool and to the filter and pump, also conceal a storage area used for pool supplies and garden tools.

Construction details

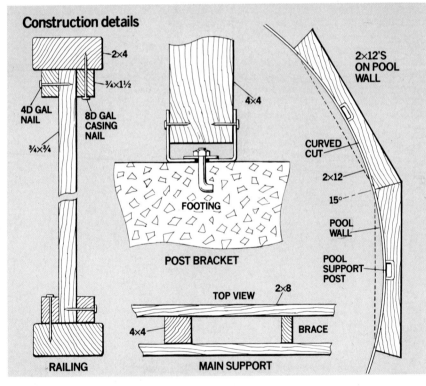

We began by leveling the ground for the 4 × 24-foot circular pool. We had to excavate 3 feet deep where our lot sloped up to the back fence.

We dug a 28-foot-diameter hole to provide working room around the pool wall. Wherever the dirt bank was more than a foot high, we built a retaining wall with corrugated steel roofing set on edge and supported by 6-foot steel stakes that were driven into the ground.

Footings. We attached the back ledger board to our 4 × 4 fence posts, which are in 30-inch concrete footings. For the corners, we used a two-man, 12-inch power auger to drill twenty-one holes 48 inches deep. (Our frost line in Nebraska is approximately 42 inches.) We filled the footings with a total of 2¼ yards of concrete and placed a ½-inch bolt in the center of each. The brackets we used to hold the 4 × 4 posts—they are known as "AB" Adjustable Port Braces and are manufactured by the Simpson Company—worked well and, in fact, saved us a lot of trouble.

These brackets served two important functions:

1. They held the posts 1 inch above the ground. That protects the posts from absorbing moisture.

2. They allowed us to move the posts about 1½ inches in any direction to line them up and make them plumb during construction.

Support structure. All the 4 × 4 posts that were to be extended above the deck for railing were made of redwood to match the deck surface. The rest of the posts, and all of the 2 × 8s underneath, are cedar.

The main support joists were built with paired 2 × 8s, which were secured to the posts with 30d galvanized nails. Joist hangers and six-way corner brackets were used wherever possible to provide additional strength. The frame required approximately 1700 board feet of lumber.

Decking. All the exposed decking wood is Construction Heart redwood. The decking boards are 2 × 6s, which were spaced apart with a 16d nail. We nailed the top down with 10d galvanized casing nails (we used 35 pounds of them). As you can see in the accompanying illustrations, the pattern of the deck boards at pool level gave us a chance to learn how to make some difficult angle cuts.

One puzzle was how to cut the pool wall itself. Kirk Olsen solved this in his design, by using 2 × 12s and cutting one side to curve with the pool. We used a reciprocating saw for these cuts. The larger size board allowed us to make a curved cut on a piece and still leave plenty of board for support. The deck top required approximately 2300 board feet of Construction Heart redwood, including the railing and sides.

The railing looks complicated, but was really quite simple to build. We used 2 × 4s. For use on the side pieces and the spokes, we ripped the 2 × 4s to ¾ inch.

Sides. Since we didn't want the pool to show, we enclosed the sides. For this, we framed in 2 × 4s and covered with 1 × 8s. There are three doors for access underneath to the filter and pump. We can also use the doors for access to storage, although the area gets quite wet at times. We built a 12 × 12-inch trap door right above our filter and mounted it on a piano hinge (the filter is cleaned through top).

In addition to the underdeck storage, we built a cabinet by extending the overhang on the house down to the deck. This cabinet holds all the tools and materials needed for pool maintenance, including chemicals, which means they can't create an odor problem in the house or garage—*Gordon Muirhead.*

bi-level outdoor living

This deck, designed to add convenient outdoor living space to a bi-level home, offers direct access to its 40-foot length from the kitchen, family room, and master bedroom. Cantilevered 8½ feet above a garden patio, the deck is a lofty vantage point with a splendid view of neighboring gardens. Yet the alternating boards forming the balustrade of the deck afford the owners complete privacy.

Supported on a wood girder by five Lally columns, the deck shades a pleasant underdeck recreation area for garden living and dining. Tables and flower shelves attached to the columns and "look-outs" (supports for hanging plants) are practical, beautiful design features.

An aluminum spiral staircase preserves the light, suspended-in-air feeling created by cantilevering: a conventional staircase would have spoiled this with a space-consuming assembly of 4 × 4 posts.

The plans shown here include all details and specifications for erecting the deck. Construction Heart grade redwood was used. Here is the construction schedule:

● Install the Lally columns approximately 12 inches deep in concrete footing. Allow three or four weeks for concrete to cure.
● Fasten girder to top plates of Lallys with ¼-inch galvanized lag screws, 3 inches long.
● Place and fasten the *bottom* of the 2 × 8 ½-inch higher than the *top* of the girder ledger to give the deck a slight pitch for run-off.
● Assemble joists and joist blocking with 2 × 8 joist hangers at the ledger and 10d hot-dipped galvanized nails at the girder. (Joists were cut from 10-foot lengths to 9 feet, 1½ inches to provide blocking of 10½ inches between joists.)
● After placing the 2 × 8 header and 2 × 4 railing posts, cut 1 × 6 bracing

into the tops of joists where shown on plan.
● Lay 2 × 6 decking with 10d buttonhead galvanized nails and assemble railing as shown.
● Assembly and erection of spiral staircase requires three or four hours, including fastening to patio and deck. The staircase is made by Columns Inc., Pearland, TX 77581.
● "Look-outs" used to support hanging plants are 18-to-25-inch long 2 × 4s capped with half-round waste from Lally column tables. They are fastened between the railing posts with wedges

under the lower 3½-inch blocking so they can be easily relocated.
● For a proper deck finish, write California Redwood Association, 1 Lombard Street, San Francisco, CA 94111 and request data sheets (4B1-4) on redwood deck finishes.

Engineering your deck

For the benefit of readers who would like to know the methods and formulas for determining the safe loading of joists, girders, and Lally columns, we include architect Carl De Groote's

computations for this unique cantilevered deck.

The deck is entirely built of redwood, the specific gravity of which is .374 to .387. The weight of a cubic foot of water is 62.425 pounds. Therefore, redwood weighs 62.425 pounds × .387 = 24.158475 pounds per cubic foot, or 2.04 pounds per board foot. (A board foot is 144 cubic inches.)

A total of 1837 board feet of redwood was used on this deck. Therefore: 1837 × 2.04 pounds = 3747.481 pounds. The size of the deck is 40 × 9½ feet = 380 square feet. Weight per square foot is 3748 ÷ 380 = 9.9 pounds or 10 pounds.

Thus the dead load (weight of the structure itself) is 10 pounds per square foot. This is a weight advantage. If fir were used (weight: 34

pounds per cubic foot), the dead load per square foot would be 14 pounds—for a total dead load of 5320 pounds. Being light in weight and high in resistance to decay and insect attack, redwood qualifies as a most desirable material for outdoor construction, such as a deck.

The cantilevered area of the deck is 4½ feet wide. A deck should be designed for a live load of 40 pounds per square foot—in addition to the 10 pounds per square foot dead load. Therefore, the load per running foot on the cantilevered section will be 4½ × 50 pounds = 225 pounds. As the floor joists are to be spaced 1 foot, center to center, this will also be the load to be supported by an overhanging joist. This load could be considered to be uniformly distributed over the 4½-

foot length of unsupported joist or could be concentrated at the free end of the deck. One man standing at the railing could equal that load.

For a 2 × 8 joist, fixed at one end and loaded at the other, the safe load in pounds is equal to the width of the joist multiplied by the square of the depth of the joist times 55.6. This total is then divided by four times the length of feet cantilevered. The 55.6 is a structural quality value. For Douglas Fir and S. L. Yellow Pine it is 66.7. For West Coast Hemlock common it is also 55.6. However, for Eastern Hemlock common it is only 44.

Thus the safe load in pounds =

$$\frac{1.5 \times 7.25 \times 55.6}{4 \times 4.5} = 243.54 \text{ pounds}$$

243.54 pounds being larger than the 225 pounds required at the free end of joist, indicates that the 2×8 joist will suffice to support the 4½-foot cantilevered section of the deck, provided the joists are spaced 12 inches, center to center.

The girder is supported by Lally columns spaced 9 feet, center to center. One half of the area on the deck between the ledger and the girder will be supported by the ledger. But because the loading could occur to a maximum at the center and front

railing of the deck, I prefer to design the girder to be qualified to support the entire deck. Therefore, the segment of deck area to be considered is 9 feet $\times$ 9.5 = 85½ square feet and 85½ square feet $\times$ 50 pounds = 4275 pounds. This is the load to be carried by the girder between Lally columns. In considering the use of a girder composed of three 2×10s, the safe load in pounds is equal to two times the width of the girder multiplied by the square of the length of the girder times 55.6. This total is then to be divided

by the span in feet, or

$$\frac{2 \times 4.5 \times 9.25 \times 55.6}{9}$$

$$= 4757 \text{ pounds}$$

Since 4757 pounds is larger than the 4275 pounds design requirement it is evident that three 2×10s will suffice. A 3½-inch Lally column with a maximum unbraced length of 8 feet can

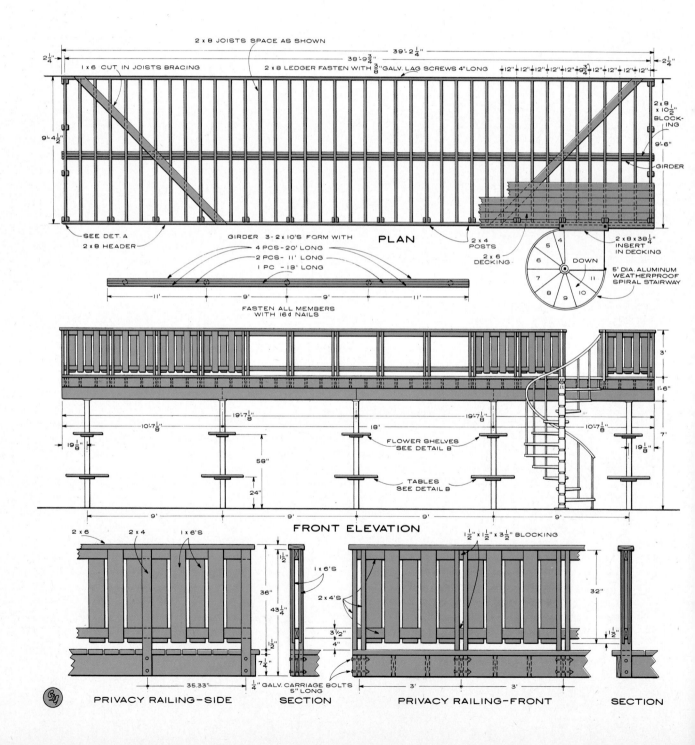

support 32,300 pounds—well in excess of what is needed. The size of the foundation and footing for each Lally column is based on the load in tons per square foot on foundation beds allowed by building codes. In New York City, for example, a load of 1 ton per square foot is allowed on soft or wet clay. A ton being 2000 pounds, an 18×18-inch footing will provide a bearing for 4500 pounds and a 2×2-foot footing will provide a bearing for 8000 pounds—*Carl De Groote* *Photos: Richard Hochman*

Raised deck is cantilevered on main girder resting on five Lally columns. This makes possible an open, under-deck patio shaded from midday sun. Redwood tables, attached directly to columns (see plan detail), are convenient for entertaining large groups. The spiral staircase comes knocked-down, can be assembled in a few hours, is attached to deck and concrete. Alternating boards of balustrade provide ventilation and ensure privacy. The deck is cantilevered at midpoint on girder made of three 2×10s. Joists are 2×8s, 12 inches on center. They are connected to 2×8 ledger on house with joist hangers. The railing posts are secured to the joists with carriage bolts.

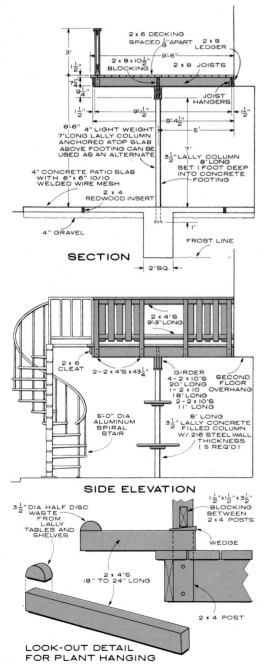

SECTION

2 x 6 DECKING SPACED ⅛" APART
2 x 8 LEDGER
2 x 8 x 10½" BLOCKING
2 x 8 JOISTS
JOIST HANGERS
4" LIGHT WEIGHT 7' LONG LALLY COLUMN ANCHORED ATOP SLAB CAN BE USED AS AN ALTERNATE
3½" LALLY COLUMN 8' LONG SET 1 FOOT DEEP INTO CONCRETE FOOTING
4" CONCRETE PATIO SLAB WITH 6" x 6" 10/10 WELDED WIRE MESH
2 x 4 REDWOOD INSERT
4" GRAVEL
FROST LINE
2' SQ.

SIDE ELEVATION

2 x 4'S 9'-3" LONG
2 x 6 CLEAT
2-2 x 4'S x 43½"
GIRDER 4-2 x 10'S 20' LONG 1 - 2 x 10 18' LONG 2-2 x 10'S 11' LONG
SECOND FLOOR OVERHANG
5'-0" DIA. ALUMINUM SPIRAL STAIR
8' LONG 3½" LALLY CONCRETE FILLED COLUMN W/.216 STEEL WALL THICKNESS (5 REQ'D)

LOOK-OUT DETAIL FOR PLANT HANGING

3½" DIA. HALF DISC WASTE FROM LALLY TABLES AND SHELVES
1½"x1½"x3½" BLOCKING BETWEEN 2 x 4 POSTS
WEDGE
2 x 4'S 18" TO 24" LONG
2 x 4 POST

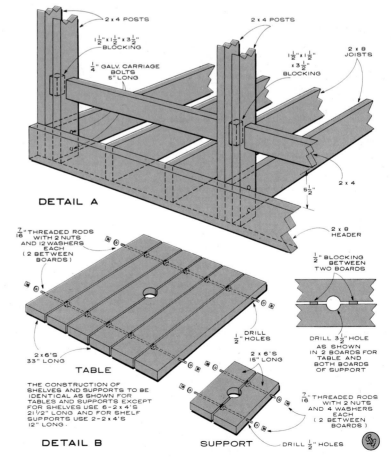

DETAIL A

2 x 4 POSTS
1½"x1½"x3½" BLOCKING
¼" GALV. CARRIAGE BOLTS 5" LONG
2 x 4 POSTS
1½"x1½" x3½" BLOCKING
2 x 8 JOISTS
2 x 4
5½"
2 x 8 HEADER

TABLE

7/16" THREADED RODS WITH 2 NUTS AND 12 WASHERS EACH (2 BETWEEN BOARDS)
½" DRILL HOLES
2 x 6'S 33" LONG

THE CONSTRUCTION OF SHELVES AND SUPPORTS TO BE IDENTICAL AS SHOWN FOR TABLES AND SUPPORTS EXCEPT FOR SHELVES USE 6 - 2 x 4'S 21½" LONG AND FOR SHELF SUPPORTS USE 2-2 x 4'S 12" LONG.

DETAIL B

½" BLOCKING BETWEEN TWO BOARDS
DRILL 3½" HOLE AS SHOWN IN 2 BOARDS FOR TABLE AND BOTH BOARDS OF SUPPORT

SUPPORT

2 x 6'S 15" LONG
7/16" THREADED RODS WITH 2 NUTS AND 4 WASHERS EACH (2 BETWEEN BOARDS)
DRILL ½" HOLES

modular deck in stages

Is the best-size deck the one you don't have time to build? Or the one you can't afford at the moment? The answer may be to start one that's modular. The design shown here is versatile enough to be built in stages or with parts of it omitted completely, the sum of any of the parts making a useful whole. Among the options are a central barbecue area with storage bins and counter tops, a sun deck, a trellis deck, and a children's play area that allows kids to play within sight of the house. All of the sections are connected by walkways or steps.

The plans were created by Duo-Fast Corp. (3702 River Road, Franklin Park, IL 60131) to demonstrate the feasibility of using its pneumatic nailer and stapler to cut assembly

Barbecue and play decks (right and below) are at extreme left and right in top view above; construction details, next page. Turn page for planter, bench, and dry-sink plans and for assembly sketch of trellis deck. Sun deck and connecting walkway are simple flat platforms.

time of large-scale projects. (Write to the company directly for information on the air-powered tools, or look into local rental.)

You can customize the modules to your house and lot: If you already have a sliding glass door to the back yard, position the modules so this access catches the barbecue deck. The plan, even if built in its entirety, still leaves enough open space for a garden, volleyball net, or small expanse of lawn— *Charles A. Miller.*

BARBECUE DECK

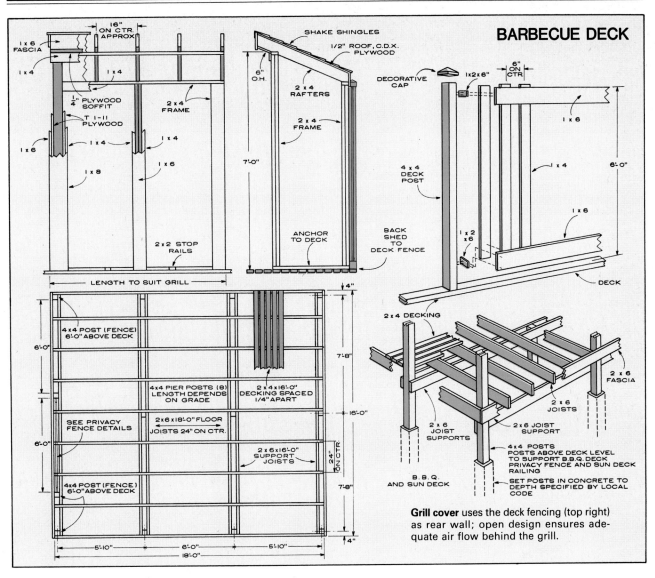

Grill cover uses the deck fencing (top right) as rear wall; open design ensures adequate air flow behind the grill.

KID'S DECK

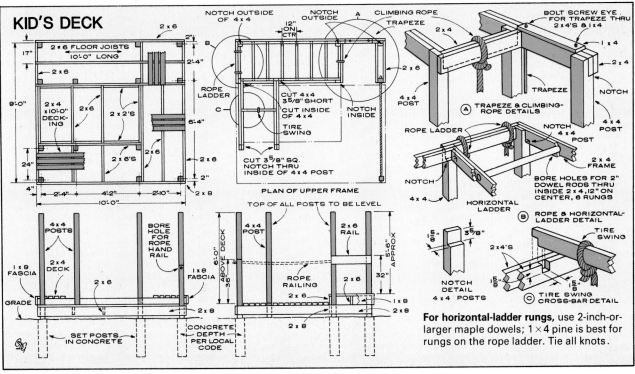

For horizontal-ladder rungs, use 2-inch-or-larger maple dowels; 1×4 pine is best for rungs on the rope ladder. Tie all knots.

TRELLIS DECK

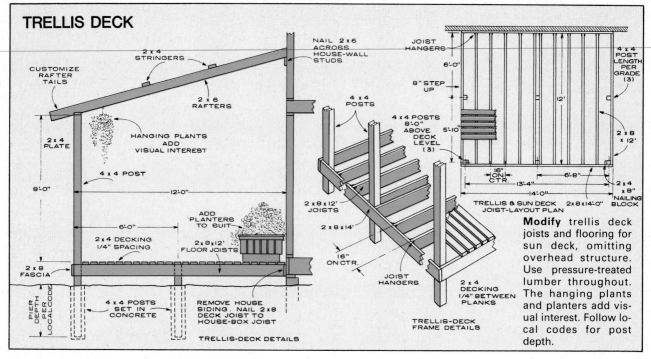

CUSTOMIZE RAFTER TAILS

2 x 4 STRINGERS

NAIL 2 x 6 ACROSS HOUSE-WALL STUDS

JOIST HANGERS

6'-0"

8" STEP UP

4 x 4 POST LENGTH PER GRADE (3)

2 x 6 RAFTERS

HANGING PLANTS ADD VISUAL INTEREST

2 x 4 PLATE

4 x 4 POST

4 x 4 POSTS 8'-0" ABOVE DECK LEVEL (3)

5'-10"

12'

2 x 8 x 12'

8'-0"

12'-0"

16" ON CTR.

13'-4"

6'-8"

2 x 4 x 8" NAILING BLOCK

6'-0"

ADD PLANTERS TO SUIT

2 x 8 x 12' JOISTS

14'-0"

TRELLIS & SUN DECK JOIST-LAYOUT PLAN

2 x 8 x 14'-0"

2 x 4 DECKING 1/4" SPACING

2 x 8 x 12' FLOOR JOISTS

2 x 8 x 14'

Modify trellis deck joists and flooring for sun deck, omitting overhead structure. Use pressure-treated lumber throughout. The hanging plants and planters add visual interest. Follow local codes for post depth.

2 x 8 FASCIA

16" ON CTR.

2 x 4 DECKING 1/4" BETWEEN PLANKS

PIER DEPTH PER LOCAL CODE

4 x 4 POSTS SET IN CONCRETE

REMOVE HOUSE SIDING. NAIL 2 x 8 DECK JOIST TO HOUSE-BOX JOIST

JOIST HANGERS

TRELLIS-DECK FRAME DETAILS

TRELLIS-DECK DETAILS

PLANTER

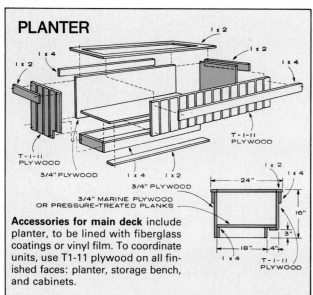

1 x 2

1 x 4

1 x 2

1 x 2

1 x 4

1 x 4

1 x 2

T-1-11 PLYWOOD

T-1-11 PLYWOOD

3/4" PLYWOOD

1 x 4   1 x 2

3/4" PLYWOOD

24"

1 x 2   1 x 4

16"

3"

18"   4"

1 x 4   T-1-11 PLYWOOD

3/4" MARINE PLYWOOD OR PRESSURE-TREATED PLANKS

Accessories for main deck include planter, to be lined with fiberglass coatings or vinyl film. To coordinate units, use T1-11 plywood on all finished faces: planter, storage bench, and cabinets.

DRY-SINK STORAGE

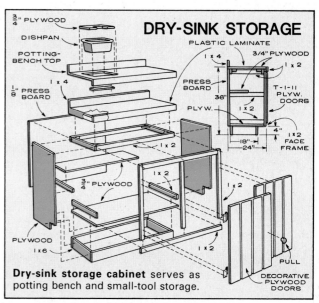

3/4" PLYWOOD

DISHPAN

POTTING-BENCH TOP

PLASTIC LAMINATE

1 x 4

3/4" PLYWOOD

1 x 2

1/8" PRESS BOARD

1 x 4

PRESS BOARD

T-1-11 PLYW. DOORS

36"

1 x 2

PLYW.

4"

3/4" PLYWOOD

1 x 2

18"   24"

1 x 2 FACE FRAME

1 x 2

1 x 2

1 x 2

PLYWOOD

1 x 6

1 x 2

PULL

DECORATIVE PLYWOOD DOORS

Dry-sink storage cabinet serves as potting bench and small-tool storage.

STORAGE BENCH

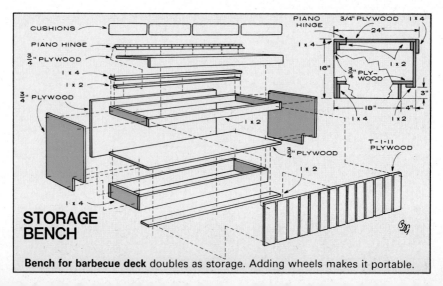

CUSHIONS

PIANO HINGE

3/4" PLYWOOD

1 x 4

1 x 2

3/4" PLYWOOD

1 x 4

PIANO HINGE

3/4" PLYWOOD   1 x 4

24"

1 x 4

1 x 2

16"

3/4" PLYWOOD

3"

18"   4"

1 x 4   1 x 2

1 x 2

3/4" PLYWOOD

T-1-11 PLYWOOD

1 x 2

1 x 4

Bench for barbecue deck doubles as storage. Adding wheels makes it portable.

folding sawbuck with limb vise

Ever try to buck a small log or tree limb with your chain saw? The log chatters, or bounces up, halting the chain and causing the clutch to slip. You can eliminate the problem by building this sawbuck, which holds lightweight wood in a vise. It's sturdy enough to handle larger logs, too. The rule of thumb: If you can lift it, the buck can hold it. It also folds flat for easy storage.

Buck parts can be fashioned from scrap 2×4s and 1×6s, three short dowels cut from an old broom handle, and forty-two 2½-inch No. 10 galvanized flathead screws. If you expect to leave the buck out in wet weather, treat the raw lumber with preservative or invest in pressure-treated lumber.

Vise parts consist of scrap 1×3s, a 3-inch long carriage bolt, two washers with a nut, two 3-foot lengths of nylon cord, and a few logs from your woodpile, to make up the lever apparatus.

Although it's okay to use a carriage bolt in the vise (because the saw won't come close to it if you cut the log to stove-wood length), don't use bolts in place of the wooden dowels as pins for crossing the sawbuck's 2×4s. Such bolts would lie in the path of a misguided saw chain. A bolt could damage the chain and perhaps even break it.

This little sawbuck doesn't take long to build. Once you're finished, you've got a buck that will not only support larger logs but also keep that lighter wood from being such a hassle and a hazard—*Neil Soderstrom.*

Sawbuck operates simply: Left foot pressing down on lever closes vise on small log, holding it firmly in place; once foot is removed, weight on other end of lever causes rig to release its grip. To open sawbuck (inset photo), set one pair of legs on ground, swing other pair outward. Center unit over lever. Note: Placement of 1×6s *A* and *B* here allows left-handed hand sawyer to place right foot inside buck under *A*. For right-handed sawyer, switch *A* and *B* heights.

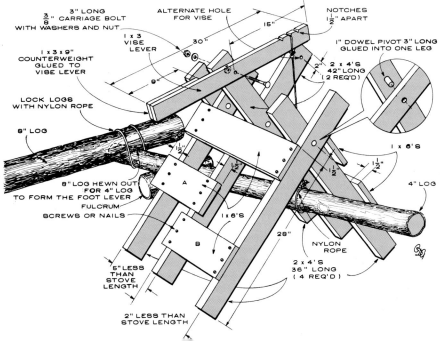

firewood shelter-fence and kindling bin

A proper place for wood and kindling adds to the pleasure of putting wood by. With this shelter-fence and kindling bin, you can forget the hassles that accompany makeshift wood covers. Here are some of the features:

● The 12-foot shelter-fence shown here will protect nearly three cords of wood—or shelter both wood and yard implements. You can increase capacity by extending the fence and increasing the number of hinged roof panels.

● The fence screens off your work area, giving privacy while sparing neighbors the temporary clutter of your cutting and splitting.

● The kindling bin will hold 2 cubic yards of bark or other kindling while it seasons.

By recycling some scrap timber and hardware, I was able to keep costs for

For access under the shelter fence or for stacking wood, you raise and prop middle panel before side panels (above). When the shelter is full of wood, simply reach through the framing to fetch it (below). Kindling bin is smaller structure at right.

the two shelters well under $100. I made my own fence posts by halving the trunk of a cedar tree that a neighbor was removing. Cedar is one of the most decay-resistant woods, and it splits easily down the center with a sledge and a few wedges. The posts should be treated with preservative to about one foot above ground level. Or you can purchase posts specially impregnated for subsoil use.

Construction

The fence posts should be sunk a minimum of 2½-feet—at least below the frost line. A horizontal notch, or dado, for 2 × 4 rails should be made with a bow saw and hatchet after the posts are erected. If you make notches on both posts before setting, it will probably mean extra digging, tamping, and hauling posts in and out repeatedly to line up the notches. Since the wood would otherwise weather rapidly, it's best to use a 2 × 4 rail that has been treated for outdoor use.

Fencing is inexpensive 1 × 2 pine mitered on one end and nailed up back-to-back to look like a picket fence. Each pair of pickets is nailed on alternate sides of the rails. By staggering the vertical pickets, you achieve an ideal compromise of screening and air circulation. And the more air circulated within the woodpile, the faster the wood will dry. Be sure to use galvanized nails for the pickets—5d or 6d should be long enough.

The covering panels are exterior-grade ¼-inch Aspenite (flakeboard) on frames of 1 × 3 pine. The middle panel overlaps the side panels to keep rain out. One 4 × 8-foot panel will shelter about a cord of wood. I trimmed mine to 7-foot lengths, anticipating heavy snow loads on the light-duty pine framing. The V-frame supporting the panels is 1 × 3 pine bolted to the frame

A sawing jig or miter box allows uniform cutting of four or five stacked slats. Here, 8-foot furring is cut to 6-foot lengths, leaving 2-footers for slatting of bin.

and joined to each other with a carriage bolt. (Were I to build another shelter-fence, I'd make the Vs narrower so that I could scoot underneath easier as the firewood is depleted.)

Shelter hinges consist of large eye screws in the frame of the panels. They rotate on a steel rod supported by eye bolts through the rail. I wrapped baling wire around the frame member holding the eye screw to prevent end-splitting. Hinges for the kindling box are just a length of pipe or a bolt through both the roof frame and the support posts (see illustration).

The kindling bin was made from scraps of odd sizes, scrap timber, and 1 × 2 furring left over from the fence construction. Vertical slats nailed inside the framing withstand the outward pressure of the kindling. The bottom of the bin is left open but is covered with masonry rubble to raise the kindling pieces off the soil. The lid is aluminum sheeting, but a flakeboard panel painted with a good weatherproof enamel would serve well too—*Neil Soderstrom.*

Use a level to keep slats true (top). Staggered slats provide good air circulation. Finished fence has birdhouses atop posts (above); they handily shed snow and rain.

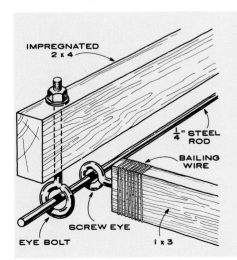

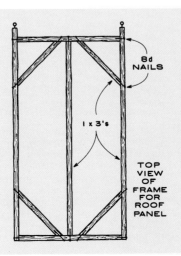

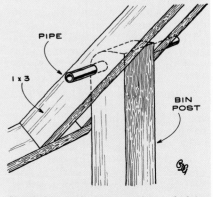

Simple hinge pin supports the roof of the kindling bin. Pins can be salvaged lengths of either pipe or just spikes.

wrap-around tree bench

Encircle your favorite tree with this hexagonal bench, and you create a cool, convenient resting place. The handsome bench is also maintenance free. It's constructed of Wolmanized, pressure-preservative-treated lumber that's totally resistant to fungus, damp rot, bacteria, and all insects, including termites. Though the bench shown here rests on a decorative stone base, the treated lumber could rest on the ground. And since Wolmanized lumber weathers to an attractive natural gray, you don't even have to paint the bench (though I did for color contrast).

The bench has two half sections, each containing three seat segments. The halves are joined at the tree. This procedure lets you work in your shop or on a level surface outside.

Each segment of the hexagon has four seat boards, two seat supports, and a pair of cross-lapped legs. That adds up to a lot of miter cuts. Such angular cutting is normally best done with a table or radial-arm saw. But for those readers who don't own such equipment, I developed a couple of simple but effective jigs. Using these special jigs, you can mass-produce the miter cuts with a portable circular saw and basic hand tools.

The dimensions shown are for a tree trunk about 12 inches in diameter, with an extra 4 inches allowed for growth. Alter the dimensions to suit your own tree taking into account its growth patterns. For a rapid grower you might want to stretch the inside diameter of the bench. And remember: Trees are wider at ground level than at seat level. I forgot this until I assembled the two bench halves around the tree and found that the rear legs were too close to the bottom of the trunk. Sawing a small triangle off the rear of each leg solved the problem.

To make mitering jig for seat boards (right), use a 2 × 4 as shown above to set exact spacing between the parallel strips. Next, measure the distance between saw blade and base edge (top right) to determine how far away from the cut line you should put the saw guide. It will be supported by lengths of 2 × 3 scrap nailed beyond the first and last spacer strips. The cut line is pencil-drawn at a 60-degree angle across the near end of the boards inserted between the spacer strips. The four seat boards were previously cut to working length.

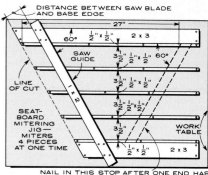

Miter-cut seat boards after attaching the saw-guide strip to the two outer support boards. The angled guide bridges the work pieces, extending about 8 inches beyond each end so the saw base is accurately guided as it enters and leaves the work. Adjust saw-blade depth so it cuts about 1/16 inch into the work board. Using this jig, cut one end of all six sets of boards before doing next step.

Nail a stop strip at a 60-degree angle across the rear end of the spacer strips. Reinsert the seat boards so the mitered ends butt up against the stop. Again using the saw guide, miter the opposite ends of the boards. If the guide is accurately set, all six sets of seat boards will assemble into a perfect hexagon.

Inviting-looking bench above is as useful as it is handsome. For precise cutting of its many parts, the author devised special jigs.

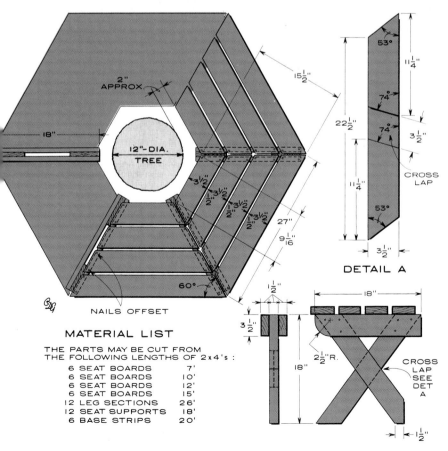

2"
APPROX.

15½"

18"

12"-DIA.
TREE

3½"
3½"
3½"
3½"
3½"
3½"
27"
9 1/16"

60°

NAILS OFFSET

53°
11¼"
22½"
74°
74°
3½"
CROSS LAP
11¼"
53°
3½"

DETAIL A

1½"
3½"
18"
2½"R.
18"
2½"
CROSS LAP SEE DET A
1½"

MATERIAL LIST

THE PARTS MAY BE CUT FROM
THE FOLLOWING LENGTHS OF 2 x 4's :

6	SEAT BOARDS	7'
6	SEAT BOARDS	10'
6	SEAT BOARDS	12'
6	SEAT BOARDS	15'
12	LEG SECTIONS	26'
12	SEAT SUPPORTS	18'
6	BASE STRIPS	20'

The plans indicate this revision, even though it doesn't show up until the last photo in the construction sequence.

To make the decorative base, first prepare border strips. Rip six lengths of 2 × 4 down to 2¾-inch width and miter the ends to 60 degrees. Next, use a pointed mason's trowel and an ice chopper to dig a 2-inch-deep, 1½-inch-wide hexagonal trench around the bench.

To do an accurate job of it, first drive a 3-inch finishing nail into the bottom of each border strip at either end. Push these spikes into the soil to hold the strips in place, then use the strips as a guide to marking trench position. Run the trowel along the edge of the strips with a sawing motion to mark and cut the turf. Remove the strips, deepen the cuts if necessary with the ice chopper, then lift out the pieces of sod. Reinsert the strips. They should project about ¾-inch above the surface of the lawn to allow neat mowing and contain the stones at the base. Varied colors and sizes of small stones are available at masonry supply houses. The stones will highlight the bench and help keep weeds down—*Ro Capotosto*.

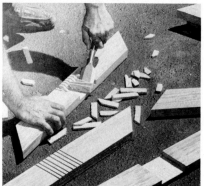

Make a third jig (bottom left) to form cross-lap joints in the legs. On an extra 2×4, nail two scraps parallel to each other, spaced so that both the saw and a 2×4 can fit between them. Make a rectangular notch in the 2×4 by cutting about eight kerfs. Clear out the waste with a chisel. To use the jig, first slide a leg piece under the guide strips, centering it against the notch in the jig. Prevent slipping by partly driving two nails through the strips and into the work. Then adjust the saw blade so it penetrates only half the leg thickness, and cut a series of kerfs through the center of the leg (top). For the end kerf cuts, run the saw against each guide. Make the inner cuts freehand about ⅜ inch apart. Note that the twelve leg pieces are identical, so there's no problem with left or right miters. Finally, use a broad chisel to knock out waste between kerfs (center). Also use the chisel to shave off high spots.

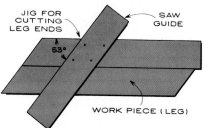

JIG FOR CUTTING LEG ENDS — SAW GUIDE — 53° — WORK PIECE (LEG)

Run a block plane over top front and rear edges of each seat piece (top) to give them a slight round. Lightly sand ends, then arrange boards in a hexagon shape (partially done here). Mark nail locations on each board, aligning marks with neighboring boards for uniformity. Nails should be slightly offset to avoid splitting seat supports. Next, make a leg-mitering jig (above) by nailing a 2×4 scrap to another scrap at a 53-degree angle. With this jig cut twelve pieces of 26-inch stock down to 22½-inch legs with parallel miters.

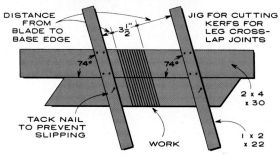

DISTANCE FROM BLADE TO BASE EDGE — 3½" — JIG FOR CUTTING KERFS FOR LEG CROSS-LAP JOINTS — 74° — 74° — TACK NAIL TO PREVENT SLIPPING — WORK — 2 × 4 x 30 — 1 x 2 x 22

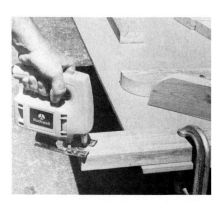

With saber saw, round off front ends of seat supports (see diagram, preceding page, for dimensions). Dense wood such as this requires coarse-set tooth blade.

Assemble bench legs first, cross-lapping pieces, then nailing a seat support to either side. Use galvanized nails and drill pilot holes to prevent splitting.

Nail seat boards to legs, aligning miters so that boards are ½ inch apart. A temporary cleat tacked to seat support helps center seat boards over legs.

Hold unfinished bench half (foreground) steady by temporary tack-nailing to the completed three-segment section. Use smooth finishing nails for this job—rough-surfaced galvanized nails are difficult to withdraw. After the final segment is inserted in place, pull out the nails to separate the two half sections.

With a level, check how evenly the bench section rests. Lawn surfaces are rarely even, so you'll have to dig away soil under some legs to obtain solid footing. If desired, put down a layer of small stones (see text). Then nail the last two sets of seat boards to the seat supports to fit the bench around the tree.

outdoor seating ideas

Does your landscaping include a place to sit? A convenient and pleasant spot to rest a bit after spading the flower bed? Extra permanent seating for outdoor entertaining? A quiet place to read a book?

Don't overlook using elements of your landscape design as a way to add seating. Put a surface on a retaining wall, for example, or if the wall is too high, divide it into two steps that can be used for sitting. Or surround a tree with a seat. Or make a seat wall of bricks or concrete blocks with caps. Garden seats can be anything from picturesque sections of log or well-placed rocks to sophisticated accessories, perhaps with colorful cushions.

Professional landscape architects almost always include seating in their plans. If you want to add outdoor seating on your property, this portfolio offers a wide range of ideas, some elaborate and some simple—*John Robinson.*

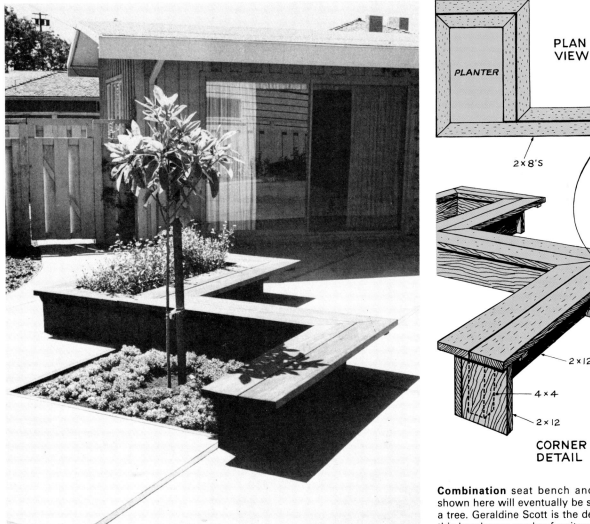

PLAN VIEW

PLANTER

2 x 8'S

4 x 4

2 x 12

4 x 4

2 x 12

CORNER DETAIL

Combination seat bench and planter shown here will eventually be shaded by a tree. Geraldine Scott is the designer of this handsome garden furniture.

The bench shown at left is light, handsome, and strong. It is relatively easy to make with a table saw. The materials (see inset) are 1-inch unsurfaced stock. The legs are sawed all on the same pattern from 1 × 4s, and a single 1 × 2 cross brace holds alignment at the bottom (see drawing). A long bolt goes all the way through to hold the legs in place. The bolt can be loosened to fold the legs for storage.

16"

16 PIECES

BOLT THROUGH,
COUNTER SINK
WASHER

This way for good seats

● Outdoor seating should be located where it will be used. A fancy wrought-iron chair, placed to show it off to best advantage, is likely to be treated like garden sculpture and seldom sat on. And don't locate seats too close to a busy pathway. A seated person should be able to stretch his legs safely, without interfering with someone trying to serve drinks or food.

● Outdoor seating should be sturdy. No one can relax on an unsteady seat. Most pieces shown in this portfolio are made with 2-inch lumber.

● Outdoor seating should be weathersafe. Pads are fine for adding comfort and a touch of color, but they should be waterproof, or easily removed for storage in bad weather.

● Make seats 16 inches off the surface—certainly not less than 14 inches. Too high, and they press painfully across the bottom of the legs; too low, and the sitter's knees stick up in front—particularly awkward when one is juggling a plate of food.

● Make the seating surface wide enough—at least a foot, preferably wider— to be comfortable for even a large person.

● Use lumber surfaced on at least one side to eliminate danger of splinters. Or do some sanding or planing on the seating side.

On the opposite page: Far left—This seat, planter, and trellis were designed as a unit by Ned Rucker. Construction around the curve is tangential. Top, near left—At the owner's request the designer, Henry Van Siegman, fashioned this seat to extend back into the planting bed. The homeowner had coil springs and pads made to fit the seat. Bottom, near left—Rustic log seat is made from large utility poles that were carefully ripped down center. The back supports are the same material, fastened from behind before the shrubs were planted. This page: Top—Seat around the tree is made of 2×12s, and it requires careful mitering. The seat surface is 2×8s with supports hidden. The paving extends under the seat. The designer is Thomas Church. Bottom left—Cantilevered seat is anchored to 4×6 timbers, which are fastened down at other end. A part of the load is taken up by the 2×4s at the bottom, which thrust against the wall. The design makes it easy to clean under the seat. Bottom right—Attractive low wall creates a raised bed for planting, although it is too low and too narrow for any lengthy sitting. For brick seat walls, figure common brick as 2¼ inches thick, about ½ inch for mortar, so five or six bricks high is right.

The Thomas Church design at top of this page is made of 2 × 12s (including seat surface). Supports are 4 × 4 posts that hold vertical pieces. In addition to providing casual seating, the low walls separate paved area from raw earth around tree. Pictured at bottom of page is a bench that provides auxiliary seating around swimming pool. Bench is a simple, box-type construction of 2 × 6s and 2 × 4s, with lid that lifts up to provide storage for items needed in maintaining the pool. Bench to fit within a curved wall (top, page opposite) is built in sections. Seat surface is 2 × 2s with pieces of 1 × 2 as dividers. Supports are 2 × 8s. Sections are slightly tangential to fit around curve (see construction details). The designer: Robert Babcock. The simple, easily portable bench shown at center of page opposite is made somewhat like a sawhorse; it can provide seating at a picnic table, can be moved out of sun or rain. Robert Babcock also designed the yard furniture shown at the bottom of the page opposite. The landscape architect added the built-in seat and planter to the railing that was required at the edge of the terrace.

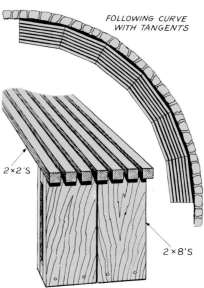

FOLLOWING CURVE
WITH TANGENTS

2×2'S

2×8'S

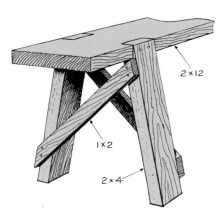

2×12

1×2

2×4

mini-catamaran

"Easy" is the word we had in mind when we designed this Mini-Cat. It's a well-mannered little craft that's easy to sail. It's also easy to build, transport, set up—and pay for. We built ours for under $350.

We also wanted the Mini-Cat to be as speedy as production boats costing five times as much. So we included design elements usually found only on much larger, new-generation cats.

The Mini-Cat is one of the few 14-footers to carry a jib. The mainsail has a high aspect ratio—its tall, thin shape is more efficient when sailing into the wind. A plastic fairing gives the mast an airfoil shape. This "wing mast" automatically pivots to face the wind on each tack. To cut drag, the halyard is inside the mast.

There's no daggerboard drag, either—the super-narrow, asymmetrical hulls are shaped to sail close to the wind and tack without a board.

You don't need any special tools or hardware to build this sophisticated little craft. What's more, all parts are from standard lumberyard stock (see diagram). Because the hulls are a simple box section without stringers, assembly is fast, and comparatively few fasteners are needed.

The sails, made from heavy-duty woven polyurethane used for tarpaulins, can be made in one afternoon—without sewing. These sturdy sails withstand high winds and are sun-resistant.

The Mini-Cat's heaviest parts are its 65-pound hulls; so one person can easily transport and assemble the craft. Once on the water, its handling qualities are quickly evident. Even beginners have found it easy to tack on a first sail. And when the wind comes up, it accelerates smartly, showing top speeds of 15 knots. That's fully competitive with far more costly production cats of the same size—*Susanne and Peter Stevenson.*

Unique mast fairing and halyard track are made of PVC pipe. Mast pivots in greased socket. Detailed construction plans, including instructions for making sails and trampoline, are $7 in U.S. money from Stevenson Projects, Dept. PS–10, Box 584, Del Mar, CA 92014. Make checks or money orders payable to Stevenson Projects.

No trailer needed—the lightweight modules are easy to car-top. The home-built cat is a good choice for those with subcompact cars—and ditto bank accounts.

Modular Mini-Cat (right) is designed for one-person setup. The 14-foot hulls are easily maneuvered into place. The trampoline (on the sand between the hulls) is preattached to the crossbars, which are simply bolted to the decks. It is then laced into place. The mast is part plastic (see inset in diagram, next page), so it's light enough for one person to step. A pull on the tiller bar kicks the rudders up for easy launching (top left). Gear and refreshments stow snugly away in twin hull hatches. Once under sail (top right), the little cat is surprisingly speedy, yet is quite docile when coming about—a tough maneuver for many cats.

concealing plywood edges

Plywood panels are a boon to the woodworker, but they have unattractive edges that can't always be concealed by joint design. Exposure occurs at front edges of case goods and shelves, and at the perimeter of tabletop sections, trays, and the like.

Some of the more traditional ways to cover edges:
● Heavy wood strips, glued and nailed in place. Standard moldings that cover both single edges and corners are available in various styles.
● A raised lip can sometimes be used. A straight strip conceals the edge; triangular strips or molding blends the lip with the slab. If you cut a rabbet in the strip you intend to use to cover the edge, one piece does the work of two.
● Edges can be concealed during construction by using solid stock inserted at corners where pieces meet (glue blocks behind corners give support). However, this requires preplanning, as inserts take up space and may change a project's dimensions.

A relatively new and easy way to conceal edges is with wood-tape products. They are suitable for use on all natural finishes.

The wood tapes come in rolls and are actually flexible bands of veneer. More and more types have come on the market, so today you can find a match for just about any species of wood used as a surface veneer on plywood. The tapes are thin enough so you can cut them with a knife, yet strong enough so you don't have to worry about breaking them.

Different types are available. Some are self-adhesive, others are applied with a white glue or contact cement. If you work with a white glue, it is best to apply thin coatings to both surfaces and wait for the glue to become tacky before placing the tape. If you work with contact cement, follow instructions on the container. You must be sure the initial placement is correct because the cement bonds immediately.

One exceptional product, so far as application is concerned, is simply called Woodtape®. It has a 1/48-inch thick veneer and a factory-applied thermo-setting adhesive. Don't let the "thermo-setting" worry you, because heat for the application is available from an ordinary electric household iron. Tapes come with a paper backing you peel off before use. They are sticky enough so the tape holds, but the bond will not be effective or permanent un-

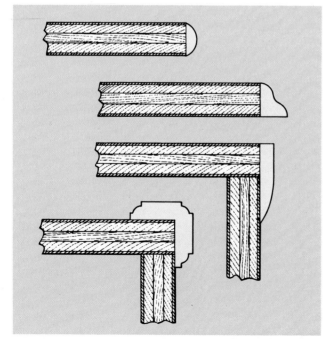 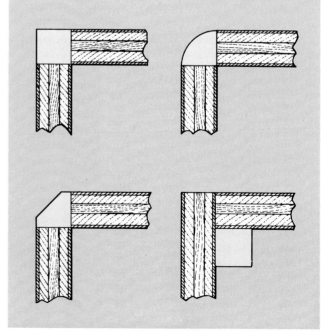

Standard moldings, such as those shown in the drawing above, left, can be used to hide single edges, as stops, or as outside corner guards. To cover the edges of plywood at the corners, you can use solid-stock inserts, as shown above, right. The inserts can be square, quarter-round, or triangular, and they can be reinforced with glue blocks.

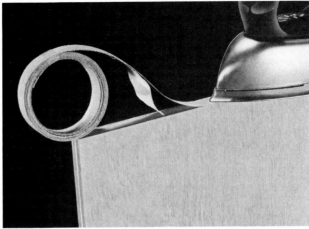

Both edges of plywood are concealed by wood strips as shown in the photos. Example on the left side of the photo requires two rabbet cuts on the wood strip and a groove in the plywood. Thermo-setting Woodtape®, which is shown in the photo directly above, is pressed on with ordinary household iron, set at 400 degrees F. Don't strip off long pieces of the paper backing in advance—remove it piecemeal as you go.

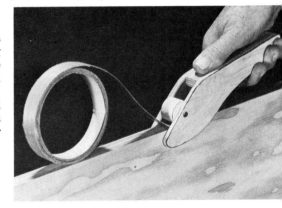

til you run over it with an iron set at 400 degrees F—about the correct setting for cotton.

Be sure the panel edges are square, smooth, and free of sawdust no matter what tape you apply. It is not necessary to fill the edges, but you should plug any large cavity that might cause a hollow to form in the tape.

Thermo-setting tape can be worked on as soon as it has cooled— contact-cement applications are ready to go right away—glue jobs require the cor-rect amount of set time. Sand the tapes as you would any veneer—being aware of their thinness. All are quite smooth, so minimum sandpaper work is needed.

Incidentally, the tapes may be used as inlay strips and, because they cut so easily with a knife or shears, for marquetry—*R. J. DeCristoforo.*

©1977 by H. P. Books, P.O. Box 5367, Tucson, AZ 85703. From *Handtool Handbook for Woodworking* by R. J. DeCristoforo.

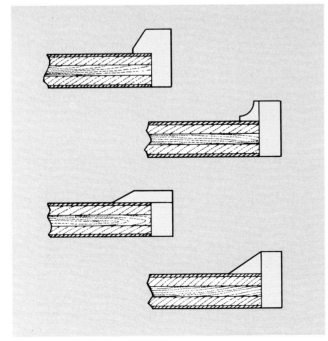

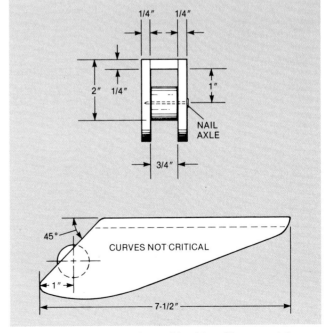

Raised lip (left, above) can be made with a triangular strip, molding two wood strips or one piece with rabbet. Shown at right are a picture of (top), and plans for (bottom) an edge-guide you can make from ¼-inch plywood to help you align wood tapes and apply pressure during gluing. The roller—it rides a nail axle—is cut from a 1-inch hardwood dowel.

calculating angles for compound miters

Recently I helped my son figure out how to calculate the cutting angles to make a planter with tapered sides. Deriving the two formulas was tough, but you'll find *using* them (with a calculator equipped for trig functions) makes it easy to calculate saw angles for cutting the sides of any structure with identical, sloping sides—picture frames, for example.

Don't let the mention of trigonometry scare you off. You don't have to be a whiz at trig to use this method. And you don't need a fancy, expensive calculator. Any low-cost model with preprogrammed trig functions (including inverse or arc functions) will do the job. A table with selected settings is provided here for those who don't have access to a suitable calculator.

To calculate saw-angle settings, you have to know the number of sides and the slope angle between any side and the vertical of the structure you're building. The number of sides determines the value of an interior angle—called angle A in the hexagonal planter illustrated. This value will be needed in the formulas that follow. The interior-angle figures below give angle-A values for regular structures with three to fifteen sides. You can

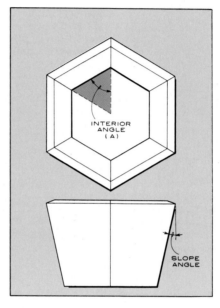

confirm these values using this formula:

$$\text{angle } A = 90° - \frac{180°}{\text{no. of sides}}$$

Practical slope

The slope angle may be anything between zero and 90 degrees, but the practical range is from about 5 to 70 degrees. Once you know the slope angle and angle A, you can calculate the arm angle and blade-tilt angle for your radial-arm saw, or the miter and tilt angles for your table saw.

Here are the two formulas used to calculate saw angles (the abbreviated forms of trig functions, such as tan for tangent, are used):

$$\tan(\text{arm angle}) = \cot(A) \times \sin(\text{slope}), \text{ and}$$
$$\sin(\text{blade-tilt angle}) = \cos(A) \times \cos(\text{slope})$$

The first equation shows that the tangent of the arm angle is equal to the cotangent of angle A times the sine of the slope angle. (Most calculators don't have a direct cotangent function, so you'll use the 1/x function on your calculator to find the reciprocal of the tangent, which is the cotangent.) The second equation says the sine of the blade-tilt angle is equal to the cosine of angle A times the cosine of the slope angle.

To calculate the saw angles, you simply plug the values of angle A and the slope angle into the right side of these equations, and solve for the arm angle and blade-tilt angle. The calculator makes this very easy. Note

	3 sides		4 sides		5 sides		6 sides		8 sides	
Slope angle	Tilt angle	Arm angle	Tilt angle	Arm angle	Tilt angle	Arm angle	Tilt angle	Arm angle	Tilt angle	Arm angle
5	59.6	8.6	44.8	5.0	35.8	3.6	29.9	2.9	22.4	2.1
10	58.5	16.7	44.1	9.9	35.4	7.2	29.5	5.7	22.1	4.1
15	56.8	24.1	43.1	14.5	34.6	10.7	28.9	8.5	21.7	6.1
20	54.5	30.6	41.6	18.9	33.5	14.0	28.0	11.2	21.1	8.1
25	51.7	36.2	39.9	22.9	32.2	17.1	26.9	13.7	20.3	9.9
30	48.6	40.9	37.8	26.6	30.6	20.0	25.7	16.1	19.4	11.7
35	45.2	44.8	35.4	29.8	28.8	22.6	24.2	18.3	18.3	13.4
40	41.6	48.1	32.8	32.7	26.8	25.0	22.5	20.4	17.1	14.9
45	37.8	50.8	30.0	35.3	24.6	27.2	20.7	22.2	15.7	16.3
50	33.8	53.0	27.0	37.5	22.2	29.1	18.8	23.9	14.2	17.6
55	29.8	54.8	23.9	39.3	19.7	30.8	16.7	25.3	12.7	18.7
60	25.7	56.3	20.7	40.9	17.1	32.2	14.5	26.6	11.0	19.7
65	21.5	57.5	17.4	42.2	14.4	33.4	12.2	27.6	9.3	20.6
70	17.2	58.4	14.0	43.2	11.6	34.3	9.9	28.5	7.5	21.3

Notes: All angles in degrees. Slope angle measured from the vertical. For table saw, substitute miter angle for arm angle.

that if you're using a table saw, you simply substitute the miter angle for the radial-arm saw's arm angle. The blade-tilt angle is the same in either case.

Before we try a sample calculation, a word is necessary about the differences in calculator models. Some use the algebraic entry system, in which operation keys ($\times$, $-$, $+$, etc.) are generally used just as they're encountered in equations. Other models, such as the Novus 4520 I used, have an entry system in which the operation key is used last. Another difference between calculator types involves the key for arc or inverse functions: Mine has an ARC key; yours may have an INV key. Both do the same thing. In the following example, the keystrokes given are those for my model, but the sequence should be easy to translate into one your calculator understands.

Now let's try a design example—the hexagonal planter box—to demonstrate the ease of calculation. From the listing of interior-angle figures that follow, you find that angle A is 60 degrees for a six-sided structure. Next, let's choose 10 degrees for the side slope. Plugging these numbers into the two formulas, we get:

$$\tan(\text{arm angle}) = \cot 60° \times \sin 10°,$$
$$\text{and}$$
$$\sin(\text{blade-tilt angle}) = \cos 60° \times \cos 10°$$

The keystroke sequence for the arm angle is:

Key step	**Result**
Press *6*, then *0*	Enters 60 (angle *A*)
Press *tan*	Calculates tan 60°
Press *1/x*	Converts tan *A* to cot *A*

Press *1*, then *0* | Enters 10 (slope angle)
Press *sin* | Calculates sin 10°
Press *x* | Multiplies prior results
Press *ARC*, then *tan* | Calculates arm angle (5.7° in this example)

For the blade-tilt angle, the sequence is:

Key step	**Result**
Press *6*, then *0*	Enters 60 (angle *A*)
Press *cos*	Calculates cos 60°
Press *1*, then *0*	Enters 10 (slope angle)
Press *cos*	Calculates cos 10°
Press *x*	Multiplies prior results
Press *ARC*, then *sin*	Calculates tilt angle (29.5° in this example)

To cut the first edge, set the arm 5.7 degrees to the right, and tilt the blade 29.5 degrees (down to the right) as shown in the series of photos. You won't be able to set tenth-of-a-degree angles, but set your saw angles as accurately as possible: Errors tend to be multiplied by the number of cuts you make. As a general rule, it is better to err on the side of larger angles, because then any gaps will show up on the inside of the piece.

Use a blade that provides a good finish cut, such as a hollow-ground combination or a crosscut blade. It is also important to cut all sides exactly the same width; the photo sequence illustrates how.

The large table provides saw angles for fourteen slope angles and five types of structures for those without calculators. But for that custom-designed planter box, hopper, or pyramid, a calculator and the formulas presented can give you the exact angles you want—*R. Joseph Ransil.*

Set and check angle settings, then make first cut (top). Flip board over, measure desired width, and mark rail (center) or set stop to assure equal widths for all cuts. Continue flipping board and aligning it on mark (bottom) till all sides are cut.

Table-saw cuts are similar. Use masking-tape marker or gauge block to assure uniform size. Press board against miter.

making tenons with a plug-cutter

When you're repairing or building chairs, tables, and other pieces of furniture, one of your hardest jobs usually is the cutting and fitting of tenon ends on rungs, back spindles, and side members. Square tenons require carefully chiseled square mortises, while dowels require holes that are accurately drilled into the end grain of the wood.

It's much easier and faster to form tenons on the workpiece itself, using a plug cutter. The plug cutter is mounted in either a lathe headstock or a drill press chuck.

With the lathe, the cutter is fixed and the workpiece is advanced into it with a milling vise and feed screw. To use a drill press, you lock the work in a drill-press vise clamped and aligned on the table and feed the cutter into the wood. While either method works well, the lathe system is easier to align because of the screw adjustments on the milling vise. Both give you accurate, round tenons quickly in any quantity you need.

After cutting the tenon to the desired depth, the waste wood is readily snicked away with a knife, or, in the case of multiple tenons on one piece, sawed free with a fine coping saw blade. If the tenon is smaller in diameter than the thickness of the wood, as is usually the case, it is easier to cut the shoulders first on a table saw. Set up the crosscut gauge and make a cut on each of the four sides.

Tenons cut in this way will fit precisely into holes drilled with a router bit. Since plug cutters come in ⅜-inch and ½-inch diameters, they match equivalent diameter router bits.

Don't forget to file or sand a flat on the tenon for glue relief. If you fail to do this, you will be trying to compress glue between the tenon end and the bottom of the hole—*E. F. Lindsley.*

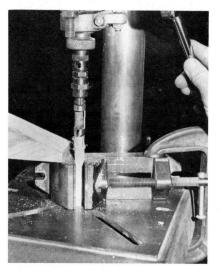

Tenons made with plug cutter are shown at top of page: Left, as formed before trimming and, next to it, trimmed and glue relief added. Others are single and double tenons formed after cutting shoulders on table saw. Plug cutter in lathe headstock rounds off ½-inch-square chair rung and leaves only slivers on corners to be cleaned up. Lower left: Same operation can be performed in a drill press, but you have to slide drill press vise around to locate it, then clamp. Lower right: Flat for glue clearance helps assembly. File or sand back almost to shoulder. Tenon holes are drilled with router bit.

router gauge for dadoes

Tack the gauge to the jig and it becomes a lot easier to cut slots in the gauge.

If laying out dadoes and cutting them with a router have always been tedious projects for you, this gauge—made of ⅛-inch hardboard—will be a handy addition to your shop. It takes just a few minutes to make, but will save you hours of figuring and improve your dadoing accuracy.

To make the gauge, you need a jig. It consists of an about 18-inch-square piece of plywood or particleboard, with a straightedge of ¼-inch stock nailed about 10 inches in from one edge (see sketch). Dimensions are not critical.

Now cut a piece of ⅛-inch hardboard 8 inches square. Measure and mark stop lines 2 inches in from each edge, as shown in the accompanying sketch. Place one side of the hardboard square firmly against your jig's straightedge and secure it with brads tacked through the outside corners. Put a ¼-inch bit in your router and set it to cut through the hardboard. Using the straightedge as a guide, cut a slot in the hardboard up to the stop line. Turning the board, repeat this process using a ⅜-inch bit, a ½-inch bit, and a ¾-inch bit, so your gauge resembles the illustration.

To use the gauge, lay out the outlines of the dadoes on the workpiece. Then at either end of the proposed cut, carefully straddle the outline with the proper slot in the gauge and mark a line at the corresponding edge of the gauge. Mark the other end of the cut in the same way and tack a straightedge on these marks. Using the straightedge to guide your router, proceed to make your cuts—*Frank H. Day.*

Using the gauge to lay out guide lines (near right) ensures accuracy of dadoes. Cutting dado with straightedge tacked to workpiece is fine for most projects (right), but with fine woods, use padded C-clamps to secure straightedge.

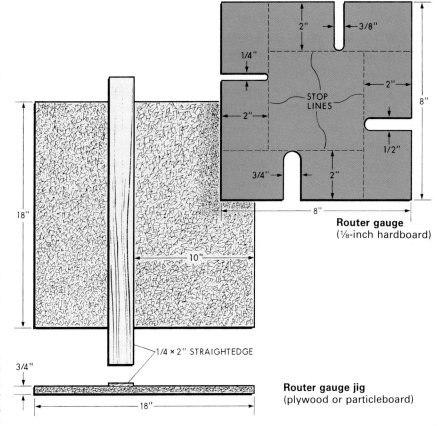

Router gauge
(⅛-inch hardboard)

Router gauge jig
(plywood or particleboard)

sharpening shop tools

Yes, razor sharp tools are dangerous—if you handle them carelessly or keep them where the young or unknowing can get at them. But many tools must be keen to do the job, and if they aren't, they'll be dangerous because you'll force them. Less than keen shop tools also increase the risk of ruining a workpiece.

Ten basics

Here are ten basics designed to guide you toward keeping your shop tools in cutting shape.
● Sharpen often. The moment a tool starts to get dull, stop and hone it, or use another until you can. It will only get duller—maybe to the point where it can't be resharpened.
● Recognize that there are some tools that it is not practical, or even reasonably possible, to resharpen. The list includes jig-saw blades, band saw blades, hacksaw blades. Simply replace them when they get dull. Some new circular saw blades are also intended to be "throwaways." Usually these are marked to that effect.
● Be aware that there are some tools you may not be able to sharpen very well—specifically, carbide-tipped tools. Many new circular saw blades have brazed-on silicon carbide tips that require grinding with a silicon carbide grinding wheel on a special grinding machine. Carbide-tipped masonry bits can be ground on a bench grinder if it has a silicon carbide wheel. A normal grinding wheel won't touch carbide.
● Consider buying simple sharpening aids. Some of the accompanying photos show relatively inexpensive ones. They pay for themselves in no time, even if you use them only once in a while. Depending on your tool collection, you may not need all those shown, but at least consider those you can use. They make tool sharpening easier and much more accurate.

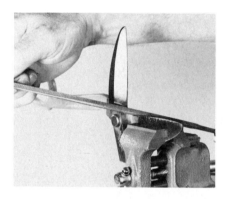
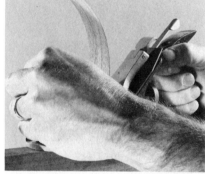

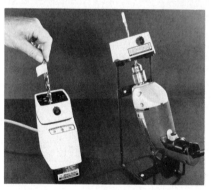

Sharp, clean-cutting tools are not only a joy to use, but they are safer and reduce chance of error and ruining work. Blades for planes (top, right) are just one of tools covered here. Many tools—such as snips at top, left—are easily sharpened by filing cutting edge carefully; rule of thumb is to duplicate maker's angle. Twist drills are among toughest tools to sharpen properly. Task is made simple with a sharpening aid. Two versions of "pencil sharpener" grinder are shown at center, right—self- and drill-powered. Little "Eclipse" sharpening aid (center, left) works well if drills are not badly worn or damaged. Sharpening is done by rolling back and forth on sheet of abrasive paper. With chisels, punches, and the like (bottom, right), chamfering the striking end to remove flare—it can chip off and hurt you—is just as important as sharpening (note two chisels at top, right in the photo, shown before head grinding). Bench grinder is a good buy. One with simple notch or angle mark on tool rest aids in hand drill grinding. At bottom, left, carbide tipped masonry drill is being sharpened with a carbide wheel.

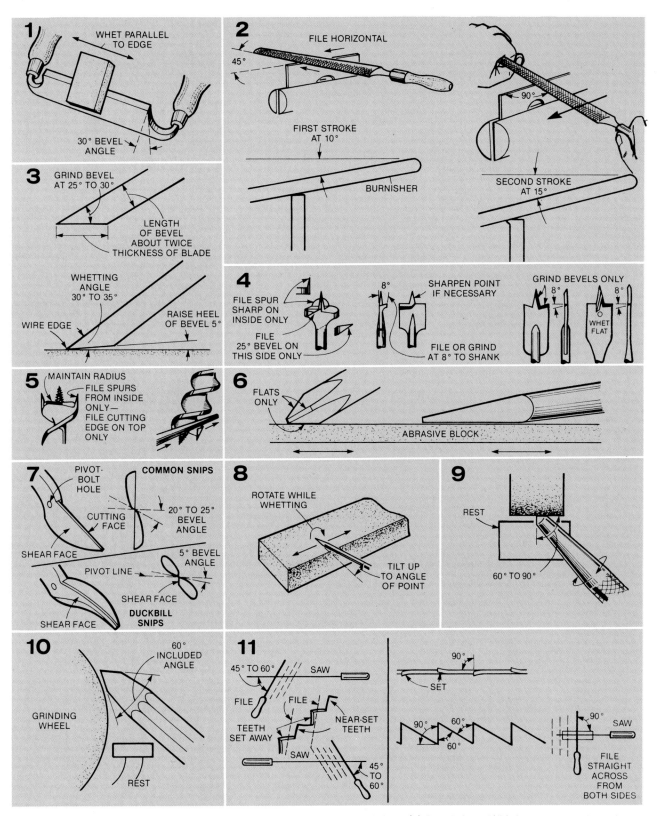

1 WHET PARALLEL TO EDGE

30° BEVEL ANGLE

2 FILE HORIZONTAL

45°

FIRST STROKE AT 10°

BURNISHER

90°

SECOND STROKE AT 15°

3 GRIND BEVEL AT 25° TO 30°

LENGTH OF BEVEL ABOUT TWICE THICKNESS OF BLADE

WHETTING ANGLE 30° TO 35°

WIRE EDGE

RAISE HEEL OF BEVEL 5°

4 FILE SPUR SHARP ON INSIDE ONLY

FILE 25° BEVEL ON THIS SIDE ONLY

8°

SHARPEN POINT IF NECESSARY

FILE OR GRIND AT 8° TO SHANK

GRIND BEVELS ONLY

8°

8°

8°

WHET FLAT

5 MAINTAIN RADIUS FILE SPURS FROM INSIDE ONLY— FILE CUTTING EDGE ON TOP ONLY

6 FLATS ONLY

ABRASIVE BLOCK

7 PIVOT-BOLT HOLE

CUTTING FACE

COMMON SNIPS

20° TO 25° BEVEL ANGLE

SHEAR FACE

PIVOT LINE

5° BEVEL ANGLE

SHEAR FACE

DUCKBILL SNIPS

SHEAR FACE

8 ROTATE WHILE WHETTING

TILT UP TO ANGLE OF POINT

9 REST

60° TO 90°

10 60° INCLUDED ANGLE

GRINDING WHEEL

REST

11 45° TO 60° SAW

FILE

FILE

TEETH SET AWAY

NEAR-SET TEETH

SAW

45° TO 60°

90°

SET

90°

60°

60°

90°

SAW

FILE STRAIGHT ACROSS FROM BOTH SIDES

Here, some tips for sharpening tools commonly found in the home workshop. **(1)** Drawknives: With hone stone, whet edge as shown; if nicked, dress with file first. **(2)** Cabinet scrapers: With file flat (A), draw it as in (B); roll over a sharp edge with 10- and 15-degree strokes with burnisher or steel bar. **(3)** Wood chisels and plane blades: Power-grind or hand-sharpen on abrasive block as in (A); whet fine secondary edge with fine abrasive block as in (B). **(4)** Spade boring bits: Pick type you have, file as shown. **(5)** Auger bits: Sharpen with file as shown. **(6)** Screwdriver blades: Hand-hone as shown. Regular screwdrivers can also be hollow ground on bench grinder. **(7)** Metal cutting snips: File bevel angle of blades as shown (normally not necessary to take snips apart). Don't file shear face or cutting face. **(8)** Awls, scribes, picks: Whet on abrasive block as shown. **(9)** Centerpunches: Maintain angle shown for job intended; rotate punch while grinding. **(10)** Cold chisels: Grind both cutting edges evenly on bench grinder. **(11)** Crosscut saws are alternately filed as in (A). Rip saw blades are filed successively, right down the blade (B).

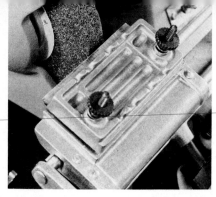

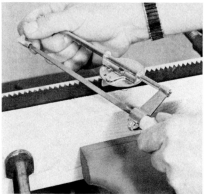
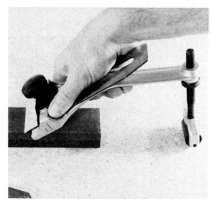

Auger bits and spade drills can be sharpened with file (top, left) small enough to clear skirt of augers. (See previous page for data on filing wood bits of different types.) Aid is important in sharpening router cutters; one shown top, center, fits on router base. Both Sears and Stanley have bench grinder attachments that fit many grinders (above right) to traverse clamped work in front of wheel. Simple, inexpensive filing aids like one shown far left, are available for both hand saw and circular saw blades. The sharpening aid shown near left positions plane blades, chisels, screwdrivers, and the like to the angle desired for honing.

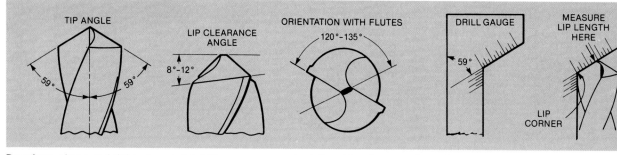

TIP ANGLE — 59° — 59°

LIP CLEARANCE ANGLE — 8°–12°

ORIENTATION WITH FLUTES — 120°–135°

DRILL GAUGE — 59°

MEASURE LIP LENGTH HERE — LIP CORNER

Drawings, above and right, show twist drill appearance and relationships that must be maintained. Of particular importance is orientation of grind with flutes. An inexpensive drill gauge is helpful in keeping the grind equal.

● Use a sharpening service when you need to. While you can, for example, sharpen either a hand-saw or circular-saw blade by hand with a file (and quite accurately with a sharpening aid), you probably have a local sharpening service that can do the job by machine much better and at a price you can't touch if you value your time at even a very low rate. The secret here is to keep spares so you won't be out of business when some of your cutting tools are being renewed. Professional sharpening is almost a must if the blade is to be reset or "gummed" (gashed deeper).

● Consider the purchase of at least an inexpensive bench grinder—or a wheel grinder accessory for another stationary tool, if you don't already own one.

● Nearly all cutting tools are made of special steels that are heat-treated and then tempered (carefully reheated to less than the hardening temperature) to give the tool edge just the right degree of hardness and toughness to do their intended jobs. Any

grinding must be done very slowly, with frequent stops to allow the steel to cool. Overheating, which destroys the temper, is to be avoided. In nearly every case, when the edge is hot enough to reach a straw color, the damage has been done. Unless the edge is retempered properly, it will never perform as well as it should.

● Get the basic equipment you need. Besides the bench grinder and perhaps some sharpening aids, needs are minimal: a good flat file; a small triangular file; perhaps a small round file; a fine/coarse abrasive block (see photo with the blade sharpening aid); a small hand abrasive stick; and a fine hone, such as an Arkansas stone.

● Know your angles. Determining the angles to which various tools must be sharpened may be a problem if you're inexperienced. A good rule of thumb: If in doubt, duplicate the angles the maker put on the tool.

● Get magnification. Unless your eyes are super good, investment in a small magnifying glass can be a big help in doing good sharpening. Why?

So you can have a good look at the cutting edge. Then you can quickly detect off-angles, nicks, flat spots, roughness, and burrs—all of which you can and should fix before the tool can be said to be really sharp.

Drill bits

Discounting the "throwaway" saw blades, chances are you use twist drill bits more than any other cutting tool. Bits are also more difficult to sharpen than other tools.

If a drill is not sharpened just right, it won't cut at all, or if it does cut it will overheat, wander off, or cut crooked. The relationship that must be maintained if the drill is to perform well is shown in one of the drawings. Your best bet is to purchase a sharpening aid for drills.

Drills can, of course, be sharpened by hand on a bench grinder without an aid. If you choose to do it that way, study the drawing to see how the point must look. Inspecting the tips of small drills is a good place to use a magnifier —*Phil McCafferty*.

how to drill big holes

Drilling holes up to ⅜ inch in diameter is usually simple enough—most of the time you can do it with twist drills. If you're working with masonry, you substitute a carbide masonry bit. But that's where the simplicity ends. Larger holes require more specialized bits.

Twist drills that you can chuck in a home-shop tool are limited in size, and their cost rises as the diameter increases. When holes get bigger than ⅜ inch, it may pay to use a bit made for the particular application. The payoff is lower cost; easier, faster drilling; and cleaner holes.

You can drill clean, accurate holes in wood using twist drills (which are designed for drilling metal)—*if* the drill is sharp, *if* you start the hole with a countersink to guide the drill, and *if* you feed the drill slowly (so hard-soft variations in the grain structure won't lure the drill off at an angle). And you must be careful to clamp the work, because a twist drill has a strong tendency to pull itself into the wood with sometimes messy results.

There's no need to risk that trouble, however. For just about every large-bore drilling application (wood, metal, or plastic), there's a specialty bit that does the job better than a twist drill, and often for less cost. Here's how to use them:

Spade bits. Chuck these bits in a portable electric drill to chew holes fast in studs and joists. The holes are rough, but it matters little in studs. In a drill press, use them to make clean-through holes—or bottomed holes, if you can tolerate the deep penetration of the point. They're virtually uncontrollable in end grain, which makes them a poor choice for drilling dowel holes, no matter how well hidden.

Auger bits. Although auger bits are usually seen with square, tapered tangs for use in a bit brace, they are also made with round shanks for chucking in a power drill. Auger bits are sized in sixteenths of an inch: a No. 8 bit is ⁸⁄₁₆, or ½, inch in diameter. The feed screw of an auger bit accurately locates the hole and helps pull the spurs into the work, but once the point breaks through, it's usually muscle power that completes the hole. Using a straight-shank auger bit on a

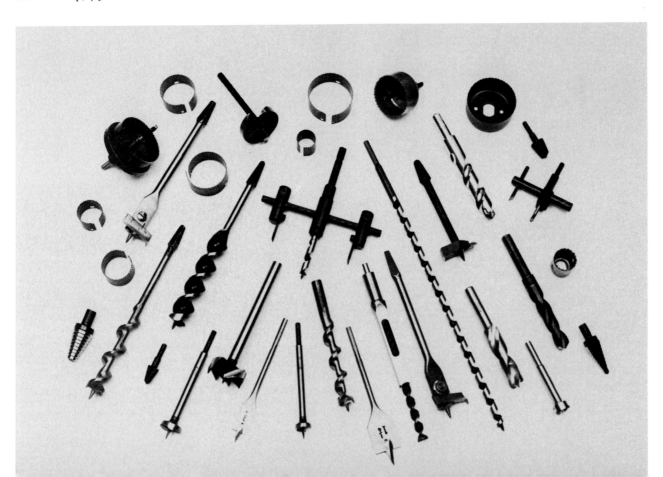

drill press or electric hand drill requires care because the feed screw can abruptly pull the bit out of control and into the work. To lessen the problem, partially file away the threads on the tapered feed screw.

Auger bits are made in several patterns. The solid-center bit is forged from bar stock. It's the strongest and least expensive. The Russell Jennings type is twisted from strip steel; it produces cleaner, more accurate holes and has better chip removal. Use long ship auger bits for boring deep, clean holes.

Power-bore bits. These bits drill clean, accurate holes in wood. They are roughly equivalent in performance to brad-point drills (below), except for the longer tip, which in many applications can be as objectionable as that on a spade bit.

Brad-point drills. The small tip (only ³⁄₁₆ inch long) on this bit starts the hole accurately and keeps it going straight. The two spurs slice through the wood, keeping the sides of the hole clean as the flutes shave out the wood. The brad-point drill is the best all-around choice for making clean holes in wood. It's best used on a drill press, but can also be used in a portable electric drill.

Forstner bits. Use these relatively expensive bits when you want to drill an especially clean hole partway through the wood without the risk of the spurs or feed screw of an auger bit (or the tip of a brad-point bit or twist drill) going deeper and possibly breaking through. The hole made by a Forstner bit is exceptionally clean

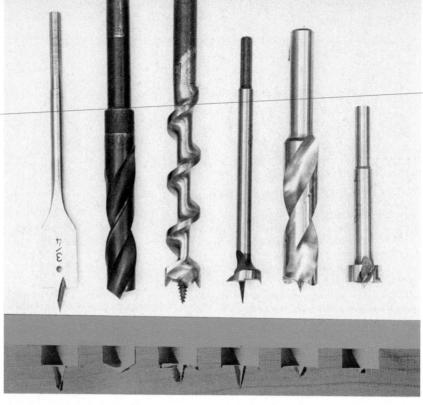

Hole profiles above are made by various bits (from left): spade, high-speed-steel twist, solid-center auger, Stanley Power Bore, brad-point, and Forstner.

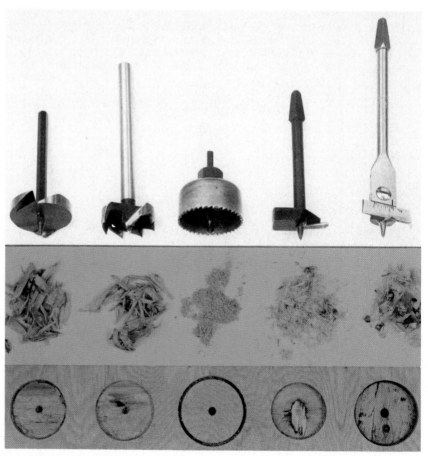

Forstner bit chucked in a drill press can drill large holes with flat bottoms diagonally in work pieces. Bit won't deflect as it enters wood.

Large-diameter wood bits, with the holes they make and the cutting debris they produce, are (from left) Stanley lock-set bit, multispur bit, hole saw and mandrel, door-lock bit, expansive bit. Holes are cut in fir plywood. Regardless of debris, all bits cut clean hole.

and has a flat bottom. Its brad point is typically only 3⁄32 inch long. Flat-bottom hole drilling is important when drilling holes for dowels in thin boards. A Forstner bit can also drill partial-diameter holes—holes slanted into the side of a board, as for screws holding a table top to an apron. The Forstner bit is for wood only; you must use a drill press; and the work, except for full-diameter straight drilling, must be firmly clamped.

Multidiameter bits. Having on hand a set of twist drills to handle all possible drilling needs in metal, plastic, and thin wood between 3⁄8 inch and 1 inch would be expensive. A plumber's tapered burring reamer is one solution to the problem, if you have a bit brace. However, maintaining accurate hole location and diameter is difficult because the tool is very fast-cutting in soft or thin materials. Arco's multidrills are single-flute tapered reamers. They are more controllable, and maintaining accurate hole location is easier. When you use the bit in a drill press, hole diameter can be accurately controlled with the depth stop.

A Unibit is the best approach to a multidiameter bit. It's a step drill with diameters in 1⁄16-inch increments. If you can work from both sides of the material, straightsided (not tapered) holes can be drilled in material up to 1⁄4 inch thick. Working from one side only, the limit is 1⁄8 inch. Even when drilling with an electric hand drill you can count steps for repetitive drilling of the same-size holes.

Expansive bits. Bits for braces are made up to No. 24 (1½ inches), and the bigger ones are expensive. A pair of expansive bits can take the place of a set of auger bits when you need holes over 5⁄8 inch in diameter, and can be used to bore holes up to 3 inches. When setting an expansive bit, always check the hole size in scrap first. Don't trust the calibration. And always be sure the cutter locking screw is as tight as you can get it.

Hole saws. There are two kinds. The inexpensive type comes with a single mandrel and a set of seven blades. The blades resemble pieces of wide hacksaw-blade stock formed into circles. The advantage of low initial cost is offset by not being able to replace worn blades individually.

In the better type of hole saw, the blade is in the form of an inverted cup with teeth around the rim. The cup mounts on a mandrel. Blades and mandrels are purchased separately, although they are sometimes available in kits. Use high-speed steel blades on wood, machinable steel, iron, brass, copper, aluminum, and most plastics (except fiberglass-reinforced and thermo-plastic). Carbide-tipped hole-saw blades cut all of the above materials, plus fiber glass-reinforced plastic, gypsum board, and plaster (but not masonry).

Both types of hole saws have twist-drill pilot drills. Use either of them in a drill press or portable electric drill. When drilling deep holes it's a good idea to back out the drill occasionally and chisel out the scrap material. The saw runs cooler with less chip clogging, and you won't have the time-consuming problem of getting the plug out of the saw cup when you're finished.

Lock-set bits. The most common need for a large-diameter wood bit around the house is for installing a new lock set on a door. You can make the large hole in the side of a door in a number of ways: You can use a hole saw; if you have a bit brace, you can use an expansive bit; if you have a lot of doors to do, you might use a special door-lock bit or a lock-set bit. If you have a drill press, you can use a hole saw, lock-set bit, or multispur bit. (Multispur bits can drill semicircular holes, just as the Forstner bit can.)

Circle cutters. Also known as fly cutters, circle cutters are used in a drill press for making large-diameter holes in metals, plastic (except fiberglass-reinforced), and wood. Work

Auger bits, from top, are double-spiral Russell Jennings, solid-center auger, solid-center auger with straight shank for drill presses, ship auger (for a flat-bottom hole).

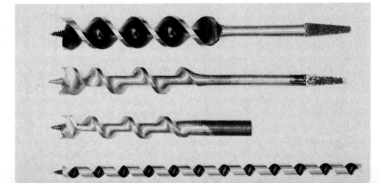

Circle cutter from Brookstone (top) cuts a 1⁄8-inch kerf, good for thick material; general's cutter (center) works best with thin material. Compass Cutter (bottom) from Hit Products works with a portable drill; its bit acts as a saw.

Multispur bit is used to drill exactly located holes in clamped wood. Performance is comparable with that of Forstner bit, except for the larger point. Bits range from ½- to 2⅛-inch diameter.

Unibit step drill bores clean holes in thin material easily. Tubing can be drilled without clamping—the bit has no tendency to walk.

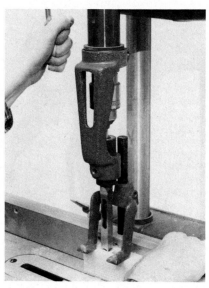

Mortising bit for drill press bores a round hole, and the chisel squares it. Considerable downward force is required for chisel to cut.

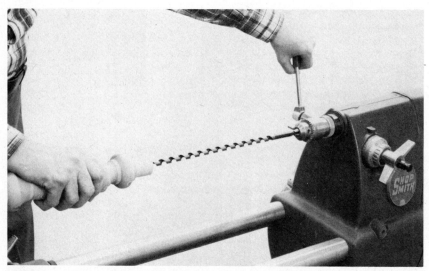

Ship auger is used for boring deep holes, such as a run for an electrical wire in a wooden lamp, above. The bit shown has a tapered feed screw, and the length of the twist section is 12 inches.

should be clamped and backed with scrap plywood. For a clean hole, cut partway from one side and finish from the other. Your drill-press speed should be slow, and the feed should be extremely slow.

A variation of the fly cutter is Hit Products' Compass Cutter, which chucks into an electric hand drill. It can cut circles up to 18 inches in diameter using a special cutter bit. The bit travels a line set by an adjustable arm tethered in the center of the cut. Use the tool for either cutting a large circle in plywood, plastic, and other similar material, or for cutting out a disc. It will cut through a 2×4.

Deep-drilling bits. Extra-long twist drills, brad-point drills, and auger bits are for boring very deep holes, such as in turned lamp parts. Of the three, the brad-point drill is the most accurate with the auger second. Use a slow drill speed and slow feed. For drilling holes in several diameters, do initial boring with a brad-point drill for accuracy, then ream with twist drills for larger diameters to save the cost of buying several expensive bits.

Mortising bits. A mortising bit drills a square hole. Use these to make mortises for tenons. You make rectangular holes by drilling overlapping square holes. The mortising bit consists of an end-cutting drill rotating inside a hollow square chisel. A special attachment clamps the chisel to the drill-press quill. The drill is chucked normally. Precise adjustment between the drill and the chisel is essential, and the chisel must be honed razor-sharp for the mortising to work at all. To drill the square hole, you have to bear down; the drill cuts a round hole and the chisel squares it. A hold-down restrains the work as you bear down and release. The advantage? You can make a lot of mortises fast, once the machinery is set up.

Drilling a big hole doesn't have to be a big project. If you have any doubt about how the hole will turn out, practice on a scrap piece of material similar to that of the project—
Thomas H. Jones.

SOME SPECIALTY-BIT SOURCES
Brookstone Co., 127 Vose Farm Road, Peterborough Road, NH 03458; **Constantine's**, 2050 Eastchester Road, Bronx, NY 10461; **Craftsman Wood Service Co.**, 1735 W. Cortland Court, Addison, IL 60101; **The Fine Tool Shops, Inc.**, 20-28 Backus Avenue, Danbury, CT 06810; **Frog Tool Co., Ltd.**, 700 W. Jackson Boulevard, Chicago, IL 60606; **Garrett Wade Co.**, 161 Avenue of the Americas, New York, NY 10013; **Hit Products, Inc.**, Box 6906, Hollywood, FL 33021; **Leichtung**, 4944 Commerce Parkway, Cleveland, OH 44128; **Sears, Roebuck & Co.** (any catalog store); **Shopsmith, Inc.**, 750 Center Drive, Vandalia, OH 45377; **Woodcraft**, 313 Montvale Avenue, Woburn, MA 01801; **The Woodworkers' Store**, 21801 Industrial Boulevard, Rogers, NM 55374.

cleaning a drill chuck

Nothing is handier than the Jacobs chuck on your electric drill—until it gets sticky from dust, dirt, and drill lubricant. Then it's twist, twist with the chuck key, over and over, just to open or close the jaws. But a thorough and complete cleanup doesn't take more than thirty minutes and, unless you've abused your chuck badly, it will work as smoothly as when it was new.

Such chucks are threaded—right-hand thread—onto or into the drill output shaft; to remove them from the drill, simply insert the key shank in a hole and whack it smartly with a hammer. There's a joker in this if your drill is the reversible type: Here, the chuck is locked by an inner screw so that it won't unscrew when the drill is running in reverse. Before removing the chuck you have to open the jaws and go in with a narrow screwdriver to remove the lock screw, which has a left-hand thread.

With the chuck off the drill, all that's necessary is to force the press-fitted outer sleeve off the central member. You can use a piece of pipe, a socket wrench, or even a hole drilled in a piece of hard wood to support the outer sleeve while you press or drive the center out. For my ¼-inch drill, a ¹⁵/₁₆-inch socket just fits the sleeve, but allows the inner member to pass inside. A vise or arbor press can be used to force the sleeve off, but if the chuck is small, a few taps with a hammer does the trick.

With the sleeve off, you can see the three sliding jaws and the split ring around them. The threads on the inner diameter of the ring mate with those in the jaws. Lift out the two halves of the ring carefully and place them in solvent (e.g., lacquer thinner), and clean, perhaps using an old toothbrush. Keep track of the top and bottom as you work, and don't turn one upside down against the other— a little care that helps avoid confusion on reassembly.

The jaws are not identical. Each is threaded a little differently so that, when acted upon by the ring threads, it moves evenly in proper relationship to the others. Should you drop the jaws on the floor, you can probably figure out how they go together by studying the threads, but for cleaning, it's a lot easier to remove one jaw at a time and then replace it exactly where it came from.

You can push the jaws out from the top with a small stick, though they might be hard to remove if the chuck is badly fouled. Wash the jaw, and especially the threads, until all surfaces are bright. Use a small, round brush to clean the hole in the chuck body. Clean each jaw and each hole, in turn, until you can slide the jaw back and forth freely in the hole over its full travel.

Here's where a bit of experience pays off. While it took just a washing out to put the jaws back into one of my chucks in free sliding condition, on another, the jaws wouldn't slide freely no matter what I did. Examination with a magnifying glass revealed minute burrs and turned edges at the end of the central member, where the jaws came out. Such burrs may be caused by using an oversize drill, loose drills that wallow and batter the chuck, and from running the chuck hard and flush against a surface. In any case, it took nothing more than some careful cleaning up of the burrs with a small hone to cure the sticking problem on my chuck.

To reassemble your chuck, slide all three jaws down until the outer ends are even in the full closed position. The two rings should drop neatly into the threads in the upper parts of the jaws. Slip the outer sleeve over the inner parts and drive or press it solidly home in the reverse of the disassembly process. Squirt a little light oil into the jaws and threaded areas and your chuck will work like new— *E. F. Lindsley.*

Reversing drills have an additional screw with left-hand thread inside chuck. Dirt may obscure screw slot until you dig it out. Chuck key keeps chuck from turning while you remove screw.

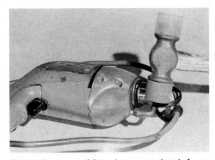

Sharp hammer blow loosens chuck from standard drill and, after screw is removed, from reversing drill, too. If it doesn't, have drill serviced. Excessive hammering may damage gear drive.

Large socket wrench supports outer sleeve, smaller one protects upper end. Hammer blow forces inner member through and out of sleeve, so inner wrench must clear OD of inner member.

Shown at bottom, are the outer sleeve, split, threaded ring, and one jaw of a chuck soaking in a solvent-filled tray. The other two jaws of the chuck are still in place— this to avoid mixing jaws up, since each is different. The hole in the inner member is brushed clean with solvent.

tapping into water pipes

It's often necessary to tap into a pipe: to install a bar sink, for example. Or maybe to add an extra outlet for your garden hose. There are many useful adapters, tap-tees, and special unions available from your hardware/home-center dealer. Knowing what connectors are available and how to use them can spell the difference between a job that's done right and one that leaks.

The accompanying photos show the more important fittings and a few of their applications. And here are spe-cific do's and don'ts about putting some of them to work for you:

PVC slip-coupling tee. This popular tee should never be connected to a hot-water-supply pipe. Polyvinyl chloride (PVC) is suited for drainage and cold pressurized water, but not for hot pressurized water. Hot water can soften PVC enough to cause a leak.

Also, plastic expands faster than metal. As soon as hot water flows, the female threads on the fitting's branch tapping expand away from a metal pipe that's threaded into them. This alone can open the joint to leaking. No plumbing code allows PVC for hot water.

Female plastic threads on a thread-ed metal pipe are not the best idea for cold water under pressure, either. This setup works well only with drainage fittings.

Celcon plastic fitting. To use this one, you simply push the chlorinated polyvinyl chloride (CPVC) or poly-butylene (PB) water-supply pipe in. The fitting holds it. No tightening is needed—or possible. It's a great idea, except that this unique fitting mate-rial cannot stand up to the chlorine almost always present in municipal water. Celcon has a weakness Cl– ions that could result in failure. Neverthe-less, this fitting is widely sold for both hot- and cold-water use.

Dielectric union. When piping sys-tems of differing metals such as gal-vanized steel and copper are joined, don't just use a male or female adapt-er. Use a dielectric union, instead.

Pipe adapters, mostly for water-supply systems: (1) Male adapter connects female pipe threads to polyethylene cold-water pipe; (2) male sweat adapter connects female pipe threads to sweat-soldered copper pipe; (3) female flare coupling connects male pipe threads to flared copper or plastic; (4) hose nipple connects female pipe threads to rubber or vinyl tubing; (5) hose-to-pipe swivel connector joins male pipe threads to male hose threads; (6) dielectric union connects male pipe threads and sweat-soldered copper; (7) Genogrip push-on adapter connects polybutylene or copper pipe to a CPVC fitting; (8) transition union adapts thermoplastic pipe to metal pipe, absorbing thermal movements (prevents leaks that would otherwise occur when two different types of pipe are heated and expand at different rates); (9) one type of quick-connector that connects washing-machine- or dishwasher-hose female threads to faucet aerator; (10) tapped bushing fits female pipe threads, presents the same, and permits a ⅛-inch pipe-thread tapping for filtered drinking fountain, etc; (11) hose nipple connects female pipe threads to female hose threads; (12) saddle-tee clamps around a drilled hole in pipe and presents female pipe threads; (13) slip-coupling tee connects sawed-off pipe ends, branches to reduced-size male pipe threads; (14) dishwasher-disposer coupling clamps over various waste pipes; (15) tailpiece for kitchen sink allows the dishwasher to drain through a common pipe.

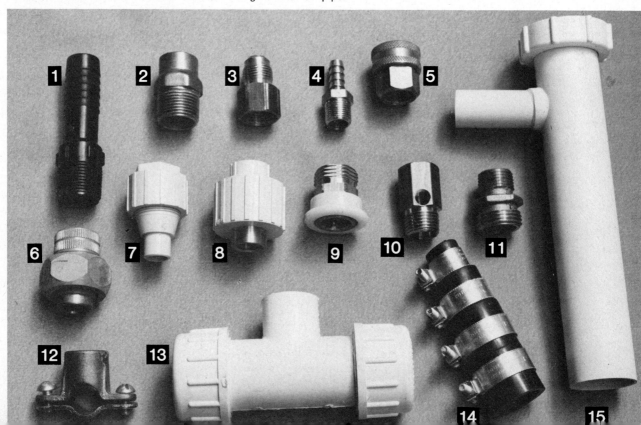

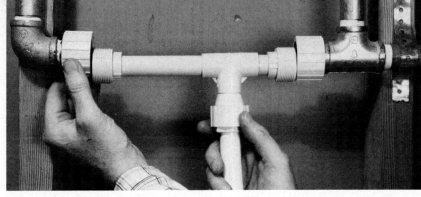

No-threading thermoplastic branch in threaded hot- or cold-water system is added by cutting and removing a length of pipe, replacing it with a pair of transition unions and a tee-containing length of CPVC pipe. Here rigid CPVC has been furtheradapted to flexible PB.

This puts an elastomeric rubber gasket between the dissimilar metals. Moreover, a plastic washer behind the union's clamping collar further breaks any electrical continuity between the two systems, greatly reducing electrolytic corrosion. The problem is especially severe with highly ionized hard water. Whether or not your plumbing code requires a dielectric union, it's a good idea to use one between dissimilar metals.

Transition union. This is another type of pipe-protecting device. It gets its name from the job it does—making the transition between metal and plastic piping systems. Every time a pressurized plastic hot-cold water-supply pipe is joined to a metal one, a transition union is necessary to prevent cracks and leaks. These are caused by the previously described differing expansion-contraction rates between metal and plastic.

One type of transition union faces off the metal and plastic across a mediating elastomeric gasket, similar to a dielectric union. However, the purpose is not to break up electrical continuity but to allow for differential thermal expansion and contraction without strain or leaking.

Transition unions are mandatory in hot-water supply systems. And they're advisable in cold-water systems. They are not needed with polyethylene (PE) pipe used outdoors and in wells. Nor are they necessary for nonpressurized applications. Simpler male adapters may be installed in these cases.

Special push-on, hand-tightened Genogrip adapters, the kind that work with copper pipe as well as plastic, are also effective transition fittings. These patented fittings are designed for PB water-supply systems, enabling adaptation between flexible PB and rigid pipes. Rubber O-rings which do the water-sealing chores, also allow for differential movement between copper and plastic.

Flare and compression fittings serve as transitions, too. They are best for small piping, such as that for humidifiers, water coolers, drinking fountains, and the like. In larger water-supply sizes they become costly. Flare and compression adapters, male or female, connect into the water-supply system. A saddle tee often has a reduced size compression fitting for copper or vinyl tubing, along with a shut-off valve. Don't expect a small saddle tee to supply water for a garden hose, though. If you want big flow, use a slip-coupling tee (for cold water) or build a no-threading, no-soldering CPVC tap-off such as that shown in

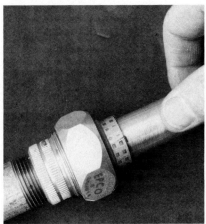

A dielectric union should be used when connecting two pipe systems made of different materials. This will prevent corrosive pipe electrolysis. The insulated jam nut joins the two systems across a nonconductive rubber-gasketed union.

No-soldering branch can be made from a copper hot- or cold-water main with a pair of push-on, hand-tightened Genogrip adapters in the copper-pipe size. Be sure they're right for copper pipe. The adapters solvent-weld to a CPVC tee.

the accompanying photos. Which one you use depends on whether your house has threaded pipes or sweat-soldered copper plumbing.

Garden-hose connectors. A wide variety of adapters and quick-couplers is available. Basic, of course, is the threaded hose bibb, but where this is not available or desired, male or female hose adapters may be used. In case you haven't tried it, a garden hose connects poorly to similar-sized ¾-inch threaded pipe, both because of the tapered pipe threads and the lack of a good gasket seal on the pipe end. Hose adapters that solve this problem are made to fit either ½-inch or ¾-inch pipes and fittings. Adapters for 1-inch hose are available, too.

From there, you can install quick-couplers for complete ease in moving hoses from one set to another. If your local plumbing-supplies dealer or hardware store doesn't have what you need, try a farm or irrigation-supplies dealer. Once every hose outlet, hose, sprinkler, and so on is fitted with a male or a female quick-connector, coupling and uncoupling take only a second. It's worthwhile where many changes are necessary.

Male and female adapters. The simplest of all, these cost little and work well where no bimetal or metal-vs.-plastic connection calls for something else. One end fits threaded pipes; the other fits sweat-soldered copper, nonpressurized plastic, rubber tube, or vinyl tube.

For use with polyethylene pipe, adapters threaded at one end and barbed at the other are available in either galvanized metal or plastic. The metal tends to corrode and the plastic may break under strain. Which to use depends on conditions. If the adapter is subject to vibration or strain, it's probably best to use galvanized. But if not, the plastic may outlast metal. You may have to shop around to find the type you want.

If you adapt to water-supply pipes correctly, the built-in, leak-free life of your home plumbing system will be kept intact. For that, it's well worth searching out the correct fitting.

If you encounter a problem not mentioned here, ask at your local home-care center for information or locate a plumbing-supply house and ask about the appropriate fittings to use—*Richard Day*.

shopping for plastic plumbing

Sorting carefully through the bins of plastic pipes and fittings in the plumbing section of a large department store near Detroit, I selected these items to put into my shopping cart: a short-shot pipe plug with half its threads unformed; a toilet flange flashed so badly that it wouldn't sit level on a floor; two drainage elbows with internal ridges that could stop up wastes; and a Y with internal recesses.

I felt sure the checkout clerk would reject the short-shot. It was so obviously unusable that anyone would have noticed it. But if she did, she said nothing. The short-shot was rung up on the cash register with all the rest of my faulty fittings.

I can't really blame the clerk, though. The bad parts should have been tossed out by the manufacturer, the distributor, or the store's buyer. But no one cared enough. If I hadn't known better, I might have used them in my home's plumbing system, later wondering what I did wrong.

This experiment confirmed my suspicions: Shopping for plastic plumbing products is like buying fruits and vegetables in a produce market. *You* must be the inspector. It is you who must reject the bananas with the soft spots, the broccoli with the too-large stems. If you pick and choose, you can end up with quality. Otherwise, you may end up with a mess.

Plastics, properly used, are great plumbing materials. They are easy to install. They don't corrode, as metal pipes do. And no flame is needed to join them.

But it isn't quite as easy as it sounds. There's a huge variety of types and materials of plastic plumbing, and choosing the proper one for a particular use can be tricky. A table gives the vital statistics for each material.

Water-supply piping can be divided into two groups: pipes able to handle hot water under pressure and cold-water-only pipes. Chlorinated poly-

vinyl chloride (CPVC) and polybutylene (PB) are the only plastics suitable for hot-water supply tubing. CPVC is a rigid plastic; PB is flexible. PB is sometimes promoted as a universal water-supply tubing, but it is at a great disadvantage where many fittings are needed. Use CPVC in applications where the lines run straight and PB where you need to run a line around corners without fittings (see table). To couple the two types of supply pipes, you'll need a brand of fittings that offers CPVC-PB adapters.

Polyvinyl chloride (PVC) and polyethylene (PE) are made as pressure lines, too, but neither material will

take hot water under pressure. Don't use them in the house. Rigid PVC and flexible PE are most useful for underground plumbing, in irrigation and sprinkler systems. PE pipe is also good in well casings.

The same two plastics are among several others used for less-demanding sewer, drainage, and drain-waste-vent piping. PVC and acrylonitrile butadiene styrene (ABS) are the most common drain-waste-vent and sewer-pipe plastics—ABS chiefly on the West Coast, PVC everywhere else. PVC drain-waste-vent and sewer pipe is made in Schedule 30 and 40 ratings. They are the same size inside, but

A gallery of faulty fittings

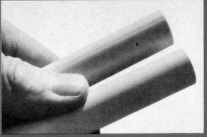

Upper CPVC water-supply pipe is marred by a full-length extrusion line. Solvent-welded joint made with it might leak.

Which spigot fitting would you choose? You'd have to fully remove price sticker off weld surface or risk bad seal.

Exterior flash harms only fitting's appearance, but it should tip you off to manufacturer's poor quality control.

Waste and drain pipes must be installed at exact slope. Badly curved pipe would create low spot, impede flow.

Close look at inside of PVC Y shows deep recesses caused by worn-out mold. This weakens fitting wall, could trap wastes.

Scuffed and flattened pipe end, caused by dragging on ground, will prevent good seal. Always inspect pipe ends.

Schedule 30 has thinner walls so it is smaller outside. It is sometimes referred to as "in-wall" pipe because a 3-inch Schedule 30 pipe (the only diameter manufactured) with its fitting hubs will slip into a standard 2 × 4 wall. To accommodate same-size Schedule 40 fitting hubs, the house wall would have to be furred out.

PE is also used for sewer and drainage pipe, or as a combination. The combination pipe may carry wastes, but not inside the house. Rubber styrene (RS) is another material often used for subsoil drainage piping.

The plastic plumbing material with the greatest heat and chemical resistance is polypropylene (PP). This plastic takes heat so well that it is used to line the insides of dishwashers. That's why it is also the material of choice for the most vulnerable part of your home plumbing system: the sink trap.

PLASTIC-PLUMBING MATERIALS

Material	Its uses	Color	Joining method
Polyethylene (PE)	Flexible cold-water pressure lines, buried or in well casings; made in various strengths; not for hot-water or indoor use; made in iron-pipe sizes	Black	Barbed clamp couplings
Polyvinyl chloride (PVC)	Rigid cold-water pressure lines buried in ground; not for hot-water use; also used for rigid drain-waste-vent, subsoil sewer and drainage piping, and fixture traps; water tube comes in iron-pipe sizes; DWV and sewer pipes come in Schedules 30 and 40; drainage pipe comes in Schedule 30	Beige; also made in many other colors	Two-step solvent welding
Chlorinated polyvinyl chloride (CPVC)	Rigid hot- and cold-water pressure lines indoors and buried; comes in copper-tube sizes	Beige; sometimes colored black or gray	Two-step solvent welding
Polybutylene (PB)	Flexible hot- and cold-water pressure lines indoors and buried; comes in copper-tube sizes	Beige; sometimes colored black or gray	Mechanical couplings
Polypropylene (PP)	Tubular drainage products for fixtures—traps, etc.; most resistant to heat and chemicals of all plastic-pipe materials	Beige	Slip-jam-nut couplings
Acrylonitrile butadiene styrene (ABS)	Rigid drain-waste-vent and sewer piping, chiefly on West Coast; also, fixture traps and slip-jam nuts	Black	One-step solvent welding
Rubber styrene (RS)	Rigid subsoil drainage piping	White, milky, or black	One-step solvent welding

Feel around insides of fitting sockets to ensure that solvent welding surfaces are smooth. Inspect threaded fittings, too.

Test-fit pipe and fitting before buying. Pipe should fit snugly so that it doesn't bottom out or fall off when it's inverted.

You will find traps made from other materials, too, including PVC and ABS. ABS suffers from intergranular stress cracking, in which contact with animal fats or vegetable oil can suddenly relieve molecular stresses created in the molding process. It happens like an earthquake on a molecular scale. PVC and PP traps are resistant to stress cracking.

If all of this doesn't have you confused already, then consider the variety of colors, sizes, and joining techniques available. Pipes and fittings may be black, beige, white, milky, or other colors. Some materials use a two-step solvent-welding process, some a one-step. Others require mechanical couplings, barbed clamps, or slip-jam nuts. Sizes are derived from the conventions used for metal pipes. Thus PE and PVC are designated like iron pipes, by their inside diameters, and PB and CPVC lines are usually measured by their outside di-

ameters, as is copper tubing. Just make sure that the pipe and fitting sizes you buy are compatible.

Now that you know the apples from oranges, keep remembering bananas when you go to the store. Be suspicious of every part you see; any of them may be defective.

Here are some of the common problems you're likely to encounter. Flash—ragged webs of superfluous plastic—is characteristic of all injection-molded plastic parts. The tiny spaces between parts of the mold fill with molten resin, which solidifies when cooled. Flash on the solvent-welding surfaces of a fitting can affect the joint. Flash on the insides of a fitting can impede flow. Quality fittings have the flash trimmed off.

The same is true for sprues, solidified fill holes where the raw plastic entered the mold. Most sprues are located on the outside of the fitting, where a small projection hurts nothing

but the appearance. Some manufacturers use inside sprues, however—a bad practice. If not ground off flush, an inside sprue forms a waste catch.

With threaded fittings, look out for poorly formed or shortened threads. Avoid using female plastic fittings with metal pipe. Plastic expands at a faster rate than the metal it's threaded onto; since it expands away from the metal, leaks are bound to result.

Beware of color variations between fittings of the same brand. These warn that the manufacturer's resin-mix control was haphazard. Be especially careful when buying white CPVC or PVC, whose natural color is beige. Manufacturers sometimes add chalk to extend the resin, but this weakens the plastic. Burned spots, dirt showing in the material, and surface roughness are other indicators of poor procedures at the factory.

Many do-it-yourselfers will still feel overwhelmed when confronted by plastic pipe in the store. Home plumbers need all the help they can get in selecting pipes, fittings, and solvent cements, but too few stores offer it. In that Detroit-area store, for example, I found pipes and fittings in white, black, gray, and beige, but no signs identified the plastics. Also, the store featured an extensive line of solvent cements varying considerably in price, but no information directed me to the proper cement for the job. If I had been a price-shopper, I might have chosen the cheapest cement and taken home a material that wouldn't work on my piping.

That's one extreme. At the other is the store with displays that clearly identify pipes and fittings and explain how they are used. The solvent cements to use with each are indicated. If you are really lucky, a clerk who understands plastic plumbing materials is available to advise.

Once you select a brand of materials, stick with it. Buying a pipe by one manufacturer, a fitting by another, and a solvent cement by a third is no route to quality plumbing. When all parts are made by the same firm you know that they were made to fit together and the solvent cement was formulated to work with them.

Don't just rely on a familiar name, though. Check for an ASTM number on each pipe and fitting. That means the part conforms to the standards established by the American Society for Testing and Materials. In water-supply piping, also look for the National Sanitation Foundation's "NSF-pw" approval, meaning parts are suitable for potable water—*Richard Day.*

PIPE OF CHOICE FOR HOT-COLD WATER SUPPLY

Use	Rigid CPVC	Flexible PB
Fixture drops (air chambers, etc.)	Best	Too many costly fittings
Fixture branches (off mains)	Best	Many costly fittings
Hot-cold mains, outdoor hose bibbs	Best for shorter runs	Best for long, fitting-free runs
Exposed in basements, utility rooms	Best	May look unworkmanlike
Service entrance, between-floor risers, garage branch	Too many fittings	Best; done with few fittings
Plumbing remodeling work behind existing walls	Difficult to run	Best; may be threaded behind walls
Around obstacles	Too many fittings	Best; done without fittings
Shower riser pipe	Best	Offers no support for shower arm
Bath/faucet pipe to spout	Not enough support; use a metal pipe	No support; use a metal pipe
At water heaters, softeners	Excellent (use proper transition fittings)	Excellent (use proper transition fittings)

how to wire a post lamp

I f you've been wanting an outdoor post lamp for both its practical and decorative attributes, want no more. You can install one, just by following the directions below.

First, decide where you'll locate your post lamp and its switch. The lamp can be placed almost anywhere, but not so close to the driveway or sidewalk that car doors may open into it or vehicles backing into the drive could inadvertently knock it over. Also, avoid property easements or installation over a septic tank or any other buried utility that may, in the future, need uncovering. The switch

would probably be most convenient at the entrance door nearest the post lamp and there is probably a switch box already there.

Before dismantling the switch box, turn off the power at the main panel box. Then take off the switch plate and carefully pull out the switch or switches. Leave them connected; at present, you just want to find out if there is a feed or source of electricity in the box.

To do this, look at the back of each switch: If there are two wires connected to a switch, it is a single-pole (S/P) switch. If there are three wire

connections, it is a three-way switch. If the box contains only one switch and there are four wires connected to it, close it up; another means of switching will be needed.

S/P switch

If there is an S/P switch in the box with two black wires connected to it, and there are two or more white or "neutral" wires spliced together in the back of the box, you have a feed. To determine which black wire is HOT, turn the switch *OFF*, uncap the white wires but leave them twisted together. Make sure the white and the black are

Floodlight assembly is installed as easily as post lamp. Use rework box in eave, push wire out from attic recess.

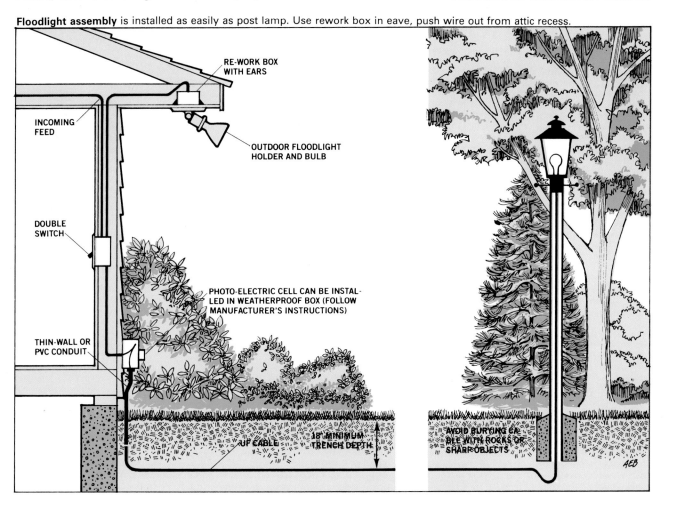

RE-WORK BOX WITH EARS

INCOMING FEED

OUTDOOR FLOODLIGHT HOLDER AND BULB

DOUBLE SWITCH

PHOTO-ELECTRIC CELL CAN BE INSTALLED IN WEATHERPROOF BOX (FOLLOW MANUFACTURER'S INSTRUCTIONS)

THIN-WALL OR PVC CONDUIT

UF CABLE   18" MINIMUM TRENCH DEPTH

AVOID BURYING CABLE WITH ROCKS OR SHARP OBJECTS

AEB

not touching and turn the power back on. Using your voltage tester, touch one probe to the white wires, and one to a terminal on the switch. Only the HOT terminal will light up the tester. Tag that wire for later reference.

Three-way switch

If there is only a three-way switch in the box, and there are two or more white wires spliced together in the back, follow the same procedure as above. This time, however, if a feed is present, the voltage tester will light up in *both* switch positions when touched between the white wires and the common terminal of the switch. If the "common" wire is HOT, tag it for later reference.

In the unlikely event there is no feed in your entrance-door switch box, and there is no other convenient place from which to switch a post lamp, you will have to use a cut-in or rework box, and add a new switch to the wall. It will then be necessary to pull a cable from a box that does have a feed—like a receptacle or another switch box—to the new box. This will require fishing the cable through the voids within the walls. You will probably need to drill holes in the basement or attic to get to wall interiors.

Buying material

After determining the location for your switch and post lamp, purchase the materials for the job. You will be using "Romex" cable or nonmetallic sheathed wiring, Type N.M. for any indoor wiring. The underground stretch will have to be Type U.F., suitable for direct burial. To determine the size to buy, inspect the fuse or circuit breaker that controls the circuit you will be utilizing. If it is a 20 amp circuit, buy #12/2 with ground; for a 15 amp circuit, buy #14/2 with ground.

You will also need a weatherproof (W/P) box, preferably with side mounting ears and a W/P blank cover; a Romex connector to fit the ½-inch hole in the back of the box; about six wire nuts; and, for use as a sleeve, a short piece of conduit and a ½-inch W/P conduit connector.

If you're adding a separate switch box, buy a regular S/P switch and switch plate. If you are going to utilize an existing switch box, buy a stack switch. In such case you will need a switch plate to accommodate the new configuration of switches installed.

Starting the job

Again, shut off the power to the switch box that you will be working with.

Remove all the switches in it to allow yourself enough room to work, but take care to mark which wires go where so you'll be able to replace them properly. Clear the bottom of the box by pushing aside the remaining wires, and knock out one of the prepunched holes in the bottom (this is where the post lamp wire will enter).

On the outside of the wall directly in line with the switch, but closer to the ground, drill a hole through to the interior. Make the hole about 1½ inches in diameter (use a hammer and chisel if necessary). The main idea is to be able to fish a wire through the wall, and still be able to cover up the hole with the W/P box. In a frame house, there will be a clear channel between the interior switch box and the exterior hole, with the exception of the insulation. In a masonry house, there will be a hollow core inside the cement blocks; it will run from the interior switch box to the exterior hole, if the two have been properly aligned, and no insulation will interfere.

Getting wire through wall

The challenge comes in getting the wire through the cavity in the wall, from the outside to the inside. In the case of the frame house, a stiff wire or

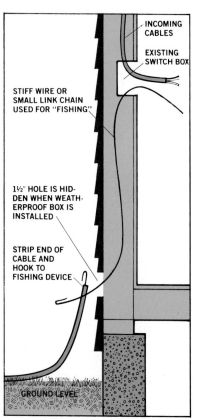

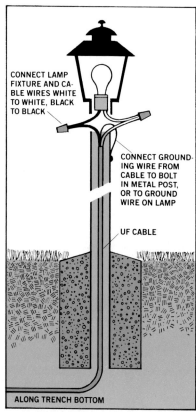

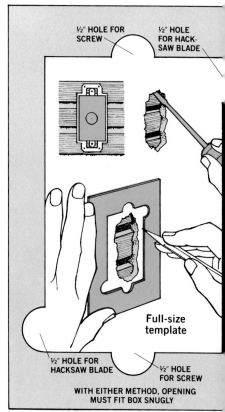

How cable is fished through wall is shown left; about a foot of cable should hang out either end. Above center: lamp wiring. Right: Two ways to install rework box. Next page: To make box with ears (No. 1) snug, simply tighten screws. Box without ears (No. 2) is retained with "Hold it" clips; they slide in next to box, tabs are bent inside. Next page, right: Typical stack switch connection

electrician's fish tape can be pushed through the knock-out hole in the switch box and forced down through the insulation to the hole on the outside of the wall. It will then be necessary to use a shorter piece of wire, and insert it in the outside hole, hook the first piece of wire and bring it to the outside, where the electric cable can then be attached and drawn up through the wall and into the switch box. In a masonry house, the same method may be tried, or a small link chain can be dropped through the knock-out hole in the switch box to the exterior hole and retrieved in the same manner as above. Attach the cable and carefully pull it up to the switch box.

Once the switch wire for the post light is inside and there is a good length hanging out of the switch box, you can cut it off on the outside of the wall, leaving about a foot exposed. Insert a Romex connector in the back of the W/P box and slide it over the exposed wire. Secure it to the wall with suitable screws.

Install a ½-inch W/P conduit connector into the threaded hole on the bottom of the W/P box and snug it up. Place the U.F. cable into a 24-inch deep trench from the W/P box to the location of the post lamp, making sure

to keep it free of kinks or sharp bends and objects. Leave enough cable at each end to make your connections. Slide a short piece of conduit over the cable to cover it from the W/P box to a point a few inches below the finish grade of the soil. Secure the conduit in the connector at the bottom of the box, and attach it to the wall using a conduit strap. Leave about a foot hanging out of the box. Cover up the cable to within a foot of the post lamp location. Slide the post over the cable, leaving about a foot hanging out of the top.

The post will probably have to be buried about 2 feet into the ground (follow the manufacturer's suggestions). Take care not to set the sharp edge of the post directly on the wire when filling in the hole with soil. Plumb the post with your level and press the soil down firmly around it.

Wire connections

When splicing household wires, simply strip the insulation back about ¾ of an inch; twist the bare ends together; snip off the very end of the twist, leaving a nice clean spiral; twist on a wire nut, and you're in business. Be sure the wire nut covers the entire bare portion of the wire.

Connecting post head

In connecting the post head to the post, connect white to white, black to black, and ground wire to ground wire. The same applies for the junction in the W/P box. As for the connection in the switch box, the white wire will splice together with the rest of the white wires. The black wire will connect to one pole on the switch, and the black wire that feeds the box will connect to the other pole on the switch. If you're using a stack switch, the feed wire connects to the side of the switch marked "Line"; the black wire from the post light connects to one of the terminals on the opposite side of the switch; and the black wire that was on the previous switch connects to the terminal right next to the post light terminal. Before you close the box, turn the power back on and carefully try each switch to see that it functions properly.

Once you have the switches working properly, turn off the power, secure the switches in the box, replace the switch cover, and turn the power back on. Enjoy your new post lamp and delight in the fact that you have just saved between $80 and $200—*Lloyd Lemons, Jr.*

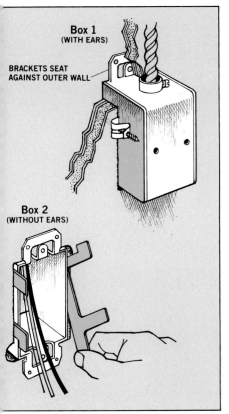

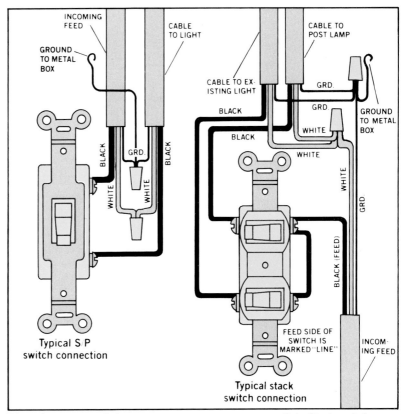

is shown at bottom; using stack switch, it's possible to have two switches occupy same space as one switch. Stack switches are available S/P over S/P or any other combination needed. Drawing at left shows typical S/P switch connection. With switch in OFF position, touch voltage tester probes from uncapped white wires to black wire. Only feed will light up tester.

secrets of drywall taping

Taping drywall is one of the toughest do-it-yourself jobs. Few homeowners have the proper tools and materials to do the job right, to say nothing of know-how or experience. And the best place to experiment is not on your living-room walls, where every crack, blister, nail hole, and tape line becomes a daily source of regret and irritation. That's why most homeowners turn to drywall professionals to do the job.

That's fine for big projects, but you may not want to call in a pro for small jobs—finishing a basement, remodeling a bath, or just repairing a damaged wall. I believe it is possible for first-time tapers to get good results using a type of drywall cement that shrinks very little and dries quickly to a hard, durable finish. Used with some other drywall-finishing products that may be unknown to non-professionals—fiberglass-mesh adhesive tape and screen-type backless sandpaper—the fast-setting compounds enable the average home handyman to finish drywall without any noticeable imperfections.

The cements I'm referring to, polyindurate compounds, aren't all that new, but manufacturers have finally overcome the lumpiness and other inconsistencies that plagued

Array of taping knives, mud pan, adhesive mesh tape, and sanding screen surrounds sacks of twenty- and ninety-minute hot mud.

past formulations. In the trade we call these compounds hot muds, because they give off mild, chemically generated heat when they are mixed. For pros, their main advantage over vinyl-type cements is that they set quickly, which means that jobs requiring two, three, or more coats can be completed in one day instead of three. That may also be an advantage for do-it-yourselfers, but hot mud's low-shrink quality is even more important. With vinyl, it is necessary to overfill drywall joints and to feather out the edges to allow for shrinkage. This is where most people go wrong, leaving too much or too little mud and slopping up the joint. You don't overfill with hot mud.

Because they have been used almost exclusively by professionals until now, hot muds have generally been available only in 25-pound bags from building-supply stores. Now some makers are packaging the powder in 4 or 5 pound packages in consumer outlets. There are many brands, including Fast Set, Quick Set, Rapid Bond, Durabond, Sudden Bond, and Sta Smooth. These are made in a variety of setup times, ranging from fifteen minutes to four hours; the most common are ninety-minute muds. There are also exterior types. Pros may use different types of mud for first-coat and finish work, but a handyman can get by perfectly well with a single mud (ninety minute or slower) mixed to differing consistencies.

The art of mixing

Proper mixing is crucial. Otherwise, you get a lumpy mud and disastrous results. Here's the right way: Put clean water in a clean bucket and pour in the powdered mud, mixing as you pour and not allowing the powder to sink to the bottom. Mix quickly to a stiff consistency, let the batch sit for a few minutes, and then remix and add water to reach the final consistency.

Different jobs require muds of different consistencies. For metal corner bead and on flat joints (both tapered board edges and full-thickness butts), you need a thick mix that clings to your taping knife. For the second coat, the mix should be softer, but not so it falls off the knife. For flattening uneven areas, taping angles (inside corners), or spotting (filling) nail holes, use a soft mix that drops off the blade. As little as one-quarter cup of water will change a bucketful of mud from one purpose to the next.

The reason a softer mix is needed for finish, spotting, and angle work is that mud stiffens considerably when worked. Those jobs involve more knife work than first-coating, in which you get rid of the mud in a single pass. When doing these jobs, make sure the mud doesn't start setting up in the pan. If a batch does begin to set prematurely, it can at first be washed away with spray from a garden hose. Later on, it will have to be chipped and scraped clean.

As important as having the proper mix is using the proper tools. Most do-it-yourselfers err by using the 6- or 8-inch knives they happen to have around. A 10-incher is the only knife to use for flat joints and spotting. For angles, you need a 6-inch knife; for feathering out on butt joints, a 14-inch may be needed. Use only straight knives, not the offset-handle contraptions with concave blades.

One other ingredient goes into a quality taping job: proper technique. That includes everything from how the knife is held to how much mud is removed from the pan to the strokes used to apply it to the wall. Journeyman tapers take months perfecting the special shoulder, back, arm, wrist, and finger motions needed to do the job effortlessly and quickly. But since do-it-yourselfers don't need a pro's speed, they can pick up sufficient skills with just a little practice.

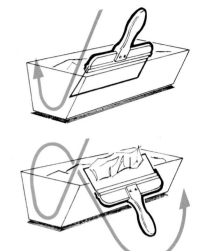

To draw mud from pan, put knife point between mud and metal (inserting whole edge risks nicked blade), dig across and up, then wipe back against pan's lip.

The first and most important motion to learn is the correct way to draw mud from the pan—what I call the figure 8. The idea is to pack the mud on the one side of your knife and leave the other side clean. Study the drawing above, and practice the repeated process of digging in, pulling up, and wiping back until it becomes second nature.

Now that your knife is loaded, you are ready to begin running angles and flats, spotting, and touching up. Each of these procedures calls for different strokes, but the knife is always held the same way: with the middle, fourth, and little fingers wrapped around the handle and the thumb and foregfinger pushing on the blade itself, not the rib. (Some tapers put the middle finger on the blade, as well.) In this position you are ready to exert either thumb or finger pressure as the task requires.

Everything I've said about technique applies equally to hot mud and vinyl. But there's one important difference. When applying vinyl cement, the knife is held at a low angle to the wallboard to allow the blade to bow and overfill the joint. With low-shrink hot mud, hold the blade so it's more nearly perpendicular to the wall, to avoid humped joints.

Now let's run through a few typical procedures. Usually, the first job you'll do on fresh drywall is taping the flats with self-sticking fiberglass mesh. This tape is superior to paper because it is thinner and doesn't require an embedding layer of mud. When taping boards that were cut to size, be sure to pound in (with the knife handle) any ragged edges that may have been left. It's crucial that the tape lie absolutely flat.

Now, using either thumb pressure and pulling up or finger pressure and pulling down (but not both), apply a layer of mud over the tape with your 10-inch knife. Then, on a second pass, scrape off the excess, leaving just enough mud to fill the holes in the mesh but with the tape visible.

On angles, conventional creased paper tape must still be used. Using a 6-inch knife, first run a thin undercoat of mud along both sides of the corner. Then cut pieces of tape to length, fold them on the crease, and press them into place. Or you can use a banjo taping machine to perform all operations at once.

After taping, you're ready for spotting—filling the small nail or screw holes created when the wall was hung. The

To begin, top center photo, string mesh tape with the thumb-to-thumb method on all flats, but don't overlap at the joint intersections. Above left, there's no pressure exchange when coating on flats; forefinger is held straight out. Above right, spotting requires a swirling motion. Here taper is finishing stroke with finger pressure after starting with thumb. Below left, use two-finger pressure for first-coating butts, leaving as little mud as possible. Below right, use 6-inch knife in angles only—for bedding, scraping, and smoothing (shown).

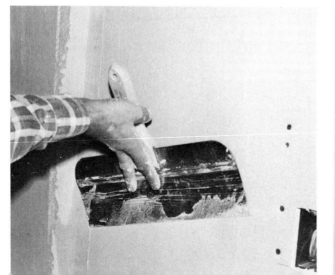
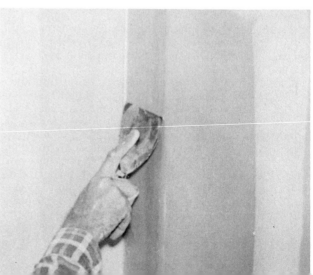

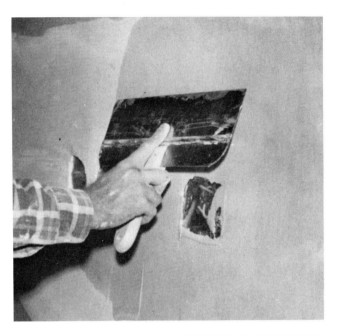

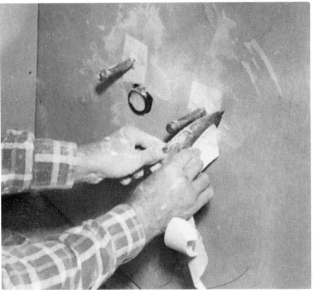

Above top, tape around electrical boxes; then coat over with mud. Above, don't use tape scraps around pipes; tear semicircles out of folded paper tape in paper-doll fashion. Below, knife blade should spring when touching up blemishes.

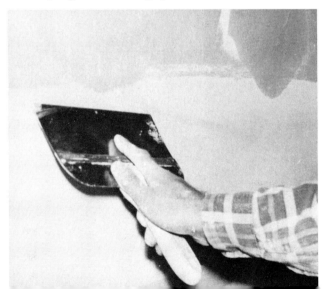

basic motion is to apply cement and wipe it clean with a single swirling motion. A pressure exchange occurs between the thumb and forefinger. If you hold the knife with your elbow turned in, then the pressure begins with the forefinger and ends with the thumb. If you begin with the elbow turned out, then the thumb pressure comes first. As you work—generally from bottom to top and from left to right (it's right to left for lefties)—look out for mud sagging in the holes. If the mixture is even slightly too thin, you must remove the mud twice from the hole, or you'll end up with hardened domes all over your walls.

Finishing for smooth walls

Once your first coat of mud is complete and fully set, the next thing to do is scrape. Anywhere the compound is not perfectly smooth—at tape edges or where the knife has left chatter or stop marks—you simply scrape lightly; the finish becomes smooth as glass. No sanding is necessary. Tape that has beaded up can be simply sliced off; its mesh network is unaffected by the loss of a piece here and there.

Now comes the finish coat. Covering the tape line on angles is easy when working with soft hot mud. Coat both sides with a 6-inch knife, then scrape off, working up from the bottom and down from the top.

Finishing butt joints requires great care to avoid leaving a noticeable hump. If the butt does not meet properly, it may have to be feathered out over a distance of 2 feet or more. After the first coat and a good scraping, slide the edge of a broad knife along the joint to locate the more severely sloped and hollow side. Fill this side slightly with your 10-inch knife. Later you'll come back for third and fourth treatments. For these, use a 14-inch knife and spread the joint out over a 28-to-30-inch width, being careful not to build up over the joint itself.

When all joints, nails, plumbing runs, and electrical boxes (see photos) have been fully smoothed out, there is one more step before you're ready to coat the wall (unnecessary if you plan to use a textured finish). Using a 14-inch knife, apply a skin coat of mud to the entire wall surface, then remove it. This leaves a frost over everything that highlights any slight imperfections that would show up when the wall is painted.

The blemishes that appear are touched up using a technique called the one-shot coat-and-clean. It's a special movement pros use to work at great speed with a minimum of return strokes to the mud pan. Although you may not need to perform with the same efficiency pros do, it may still be useful to know the proper method. The key to the technique is getting a shot of mud properly centered on the blade edge. This is done by loading the knife in the usual way, then cutting off one corner and shaking the rest down to the center. The one-shot uses the natural spring of the blade to cover a blemish and scrape the tool clean all in one motion. With finger pressure applied, push down on the blade and make it bend back, then as you pull away, elevate the handle until the blade springs back. I've seen skilled journeymen do touch-ups with this technique for several minutes without once hitting knife to pan and without leaving any chaff behind.

When the touch-up is complete, remove the skin coat. This is the only time I ever sand, using 150 paper to lightly brush down the entire surface (the backless-type sandpaper pictured on the opening page is a great boon because it never clogs). Then sweep off the dust with a soft broom and the wall is ready to be primed and painted. My favorite paint treatment for a smooth wall is two coats of semigloss latex enamel over a thinned primer coat —*Mark Lee Due.*

installing a deadbolt lock

The first step in securing your home from burglars is to make sure all entrance doors are solid and reinforced with secure locks. Of course, nothing is 100 percent secure—an intruder may smash a window, cut around a lock, or pry a door away from a jamb. In such instances, neither the finest lock, nor the most carefully selected door is any help. Still you can make it as difficult as possible for an intruder to enter. Installing deadbolt locks on doors that now have only key-in-knob locks will be a step in the right direction.

The accompanying photos and captions show how to install a mortised cylinder deadbolt lock. The lock used in this installation (The Barrier by Lori) has a 1-inch throw of hardened steel to prevent saw-through. The lock is built around a solid cast cylinder that threads into a housing, so installation varies slightly from the method used for deadbolt locks held in place by screws. (It requires one extra step and an Allen wrench—generally supplied with hardware—to join the bolt block with the cylinder housing.)

Standard tools needed for installing deadbolt locks mortised into a door are: hole saw, electric drill, ¼- and 1-inch drill bits, chisel, and punch. Before installing any lock, open the package and check that all parts have been included—*E. D. Cormier.*

Steps in installing deadbolt lock: (1) Measure your door—this is important because door thickness (1⅜ or 1¾ inches) determines center location and "backset" position of lock. **(2)** Using template supplied with lock (template must have same backset as lock you purchased), position and mark center of holes to be drilled. Center location can be marked with pencil, but hole made with punch serves as better guide for pilot drill. To insure accuracy, use small, adjustable carpenter's square. **(3)** Drill small pilot hole with 1¼-inch bit—hole must be at 90-degree angle with door and parallel to floor. Then, depending on the lock, use a 2 or 2⅛-inch diameter hole saw to drill about half way through door. Stop as often as necessary to remove wood chips. **(4)** Do not attempt to saw all the way through from one side of the door, or the hole won't be straight and will have ragged edges. After pilot bit penetrates the back side of door, drill from opposite side of door, as shown, until you have a large hole that goes completely through the door. Caution: Installation instructions vary between models and makes. Be sure to check the guidelines provided here against the instructions that were supplied with your lock. **(5)** A 1-inch flatbit, or smaller hole saw can be used to drill the second hole in the edge of the door for the deadbolt. This will penetrate through to the main hole. Chisel is then used to smooth off the excess wood, including any that protrudes in the middle of the cylinder hole and that could impair a smooth

fit. **(6)** Insert the bolt assembly with the front through the 1-inch diameter hole; mark outline as shown. **(7)** Use sharp chisel to cut the outline to the proper depth. Remove the wood in thin, lean chips, working from the hole to the top and bottom with alternate cuts. For a neat job, end with a clean break at the cut line. **(8)** Insert the cylinder housing assembly into the large hole. Then insert the bolt housing into the 1-inch hole and push it into cylinder housing. **(9)** Insert and tighten the two cylinder housing screws. **(10)** Slip the key into the cylinder. Pull it out one notch as shown in the photo; then screw cylinder into housing in door until it stops (this step varies from lock to lock; in this case, since you are using the key almost like a wrench, you must be very careful not to damage it.) Now line up the cylinder with the door—

(Continued)

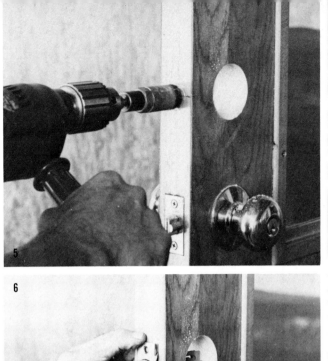

5

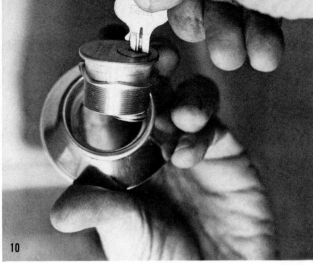

10

6

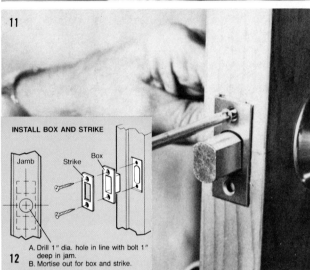

11

INSTALL BOX AND STRIKE

Jamb — Strike — Box

A. Drill 1" dia. hole in line with bolt 1" deep in jam.
B. Mortise out for box and strike.

12

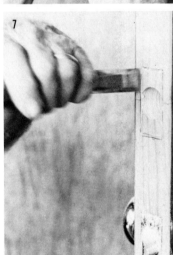

7

13

8

9

the keyhole should be in correct position—and insert an Allen wrench along the side of the bolt until it engages the cylinder set screw; tightening the set screw will join bolt block and cylinder housing. **(11)** After tightening inner cylinder and plate with Allen wrench, attach the two front screws. While tightening, check lock to insure smooth operation. Now attach the inside trim with the supplied machine screws. If your door has glass panels, you may want to install a second cylinder, rather than a turn knob. **(12)** Once the box and the strike are installed, the job is complete. Drill a 1-inch diameter hole, 1-inch deep, into door jamb and mortise out for box and strike, as shown. **(13)** Easy way to determine the location of hole for deadbolt is to coat the end of the bolt with lipstick or stamp-pad ink. With the bolt in, close the door and turn the key to throw the bolt against the jamb to mark the location of the strike.

tuning up your mower

All the spring mower-tuneup instructions you've ever read probably say the same old things:

- Change the oil.
- Put in fresh gasoline.
- Install a new spark plug.
- Clean the air filter.

After that, if you trust what you read, you just sharpen the blade, oil the wheels, and away you go—maybe.

What if the engine refuses to cooperate? There's nothing more frustrating than spending hours trying to start a balky mower. To avoid such a hassle, here's a quick series of steps that'll get your mower going again, or at least prove that a trip to the local repair shop is really necessary—*E. F. Lindsley.*

1. Whether your mower is a two- or four-cycle model, it has to have ignition, fuel, and compression to run. First check for ignition by removing spark plug, shielding it from bright light, and pulling starter rope with throttle in the "run position. Look for strong, hot spark. The clothespin shown is one way to hold a plug wire to avoid a high-voltage jolt. If you have no spark, consult your service manual for repair or take the mower into a shop. If you see a good spark, go to the next step.

2. Remove the air cleaner and test the governor and throttle linkage for free movement. Open and close the throttle to make certain the spring pulls the throttle wide open in the "start" position. Check the connecting wire and choke action if manually linked. If the choke is automatic (as shown here), put your finger in the carburetor throat to test for free movement of the plate. A gummed-up choke plate may stick (and indicates that the carburetor should probably be cleaned).

3. Spray a small amount of starting fluid into the carburetor throat and immediately pull the starter rope. The engine may now start. A burst of full-speed operation can clear a slightly gummed carburetor; if it does, you can start your work. But if the engine repeatedly starts and runs only a few seconds, the carburetor might be clogged, and it may have to be cleaned in a professional shop. But before you resort to that, check for other causes—proceed to the next step.

4. If the carburetor seems clogged, inspect the needle valve for gum (your carburetor may have more than one needle valve). Before removing a needle valve first screw it down gently and count the number of turns (and fractions of turns) it takes to seat, so you'll know where to reset it later. Remove the valve and look it over closely. Brown or gummy deposits on the tip may be a sign that a complete carburetor cleaning is needed. Clean the *valve,* then go on to the next step.

5. Squirt the carburetor cleaner directly into the needle jet. Let the cleaner soak for a minute and follow with a flushing shot. Replace the needle valve(s) at the original setting(s), squirt a small amount of starting fluid into the carburetor throat again, and try to start the engine once more. Usually it will clear its throat, cough a bit, then run—and you can get on with mowing. But if the machine refuses to crank up now, you may have some other problems. Proceed to the next step.

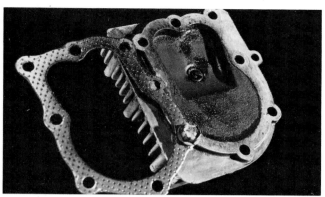

8. A typical bad gasket from a small mower engine shows dark areas (center, top) where gases were leaking. Matching area is visible on the head. Tightening the head down onto a gasket in this condition may get you going, but a new gasket, properly torqued, is the only real fix. If you have the tools and a reasonable amount of skill, you can replace the gasket yourself. In any case, do not overtighten any of the bolts; doing so will distort the head and cylinder.

6. Erratic starting or stuttering while running may be a sign that this tiny magneto ground wire on the stop switch (part of throttle bracket) is grounding out. Dirt, grease, or moisture can do it. Slip the wire out of its clip and push it free of contact with all metal. (Disconnecting this wire deactivates all interlocks; if you start your engine with it out, you must short the plug with a screwdriver to stop.) If you get spark after freeing the wire, clean and check the sliding parts of the switch.

9. A two-stroke engine must have compression in both its crankcase and its combustion chamber. The collection of oily dust on this air cleaner, plus other deposits around the carburetor, indicates that the crankshaft seals are leaking. Seals, which are vital to compression, are replaceable, but that's a shop job. Another problem in two-stroke engines is that the oil-and-gasoline mix tends to foul spark plugs easily. A plug that has its gap bridged with carbon cannot start an engine.

7. A leaking head gasket reduces compression, a common reason for hard starting. One method of checking is to apply gentle pressure against the head bolts. They should feel tight. If you find two or three that seem loose compared with others, especially in the exhaust-port area, it's almost certain your head gasket is leaking. It's not a good idea to rely on a compression gauge because you can get a false impression of what the true compression is—it may vary with each tug on the rope.

10. In a typical two-stroke engine, these six thin, flat reed valves admit fuel and air from the carburetor to the crankcase. Even fragments of grass holding one or two open will prevent the engine from starting. It always pays to check for fouled reed valves if you have the tools and experience to remove the carburetor for access. Valves shown are beginning to corrode. If there are other problems in the engine, such as bad bearings and seals (photo 9), it's probably not worth repairing.

electronic jigsaws

Using Black & Decker's two new electronic jigsaws is a bit like driving a car with turbocharger. The more load you put on these tools, the more current their electronic circuits will feed to the motors. The result is a great feeling of power. It's hard to slow the saws down, even when cutting 2-inch fir.

That power is nice to have, but it also takes some getting used to. Since these tools don't bog down, there's a great temptation to force them. And when you force jigsaws, you break blades. Even without electronics, these are top-of-the-line jigsaws. The 7566—the bigger of the two—is especially impressive. Its main features include:

● A 1-inch stroke for fast cutting and long blade life.
● A reversible shoe. Turn it around and the blade fits through a narrow slot that minimizes chipping when cutting brittle stock such as plywood.
● Automatic scrolling. You mount the blade in a special clamp on the trailing edge of the reciprocating shaft, then unlock the scrolling head. The blade is now free to "caster" and will automatically pivot as you push the saw

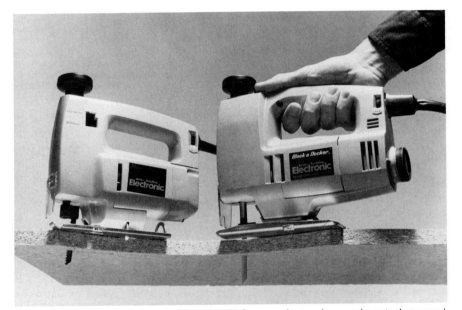

Black & Decker models 7578 (left) and 7566 feature electronic speed control, manual and automatic scrolling, sawdust blowers, and double insulation. Note that the blade on the 7578 is in the auto-scrolling position, slightly to the rear of the reciprocating shaft. In that position, the blade is free to caster.

Electronics for the 7566 model are contained on this printed-circuit board. Two solder joints and three clip connectors wire it into the jigsaw.

in the desired direction. You can also scroll manually, guiding the blade by turning the scroll knob atop the tool. Personally, I get better control this way.

Which to choose?

The smaller saw, the 7578, also has the scrolling feature. It has a shorter stroke than the 7566, but its higher stroke speed helps compensate for that to a degree. And I found its combination speed-control/thumb switch more convenient than the big saw's

separate speed dial and trigger switch. This, plus the saw's smaller size and lighter weight, would make it my first choice for doing intricate scrollwork. If, on the other hand, I wanted a jigsaw that could handle heavy, straight cuts as well as curves, I would choose the extra capacity of the 7566.

The 7578 sells for around $50, that's about 30 percent more than a roughly similar B&D saw without electronics. The 7566 is about $85. B & D makes no other saw with as much power or as many features—*A.J. Hand.*

SPECIFICATIONS B & D's ELECTRONIC JIGSAWS

Model	Maximum hp	Shoe type	Speed range (spm)	Capacity	Bearings	Approx. price
7578	1/3	Tilting	0—3200	2-inch softwood; 1-inch hardwood	Sleeve	$50
7566	1/2	Tilting, reversing	1300—2800	4-inch softwood; 2-inch hardwood	Ball, sleeve, and roller	$85

little benches for every shop

If the portable workbench lives up to its potential, it will be where you need it most, helping out with almost every job you tackle around the home. Otherwise, it could spend its entire life hanging on the wall.

After testing a few of these benches, I'm happy to report they live up to their potential. And a stable of new accessories makes them even more useful. In fact, if you add a few simple modifications and accessories that I developed while testing these tools, you'll find they qualify as genuine overachievers.

The benches I checked out were:
● The Dual 29-inch Workmate by Black & Decker, plus an exact duplicate sold by Sears under the Craftsman name.
● Wen's All Bench, another 29-inch model.
● Hirsh's Foldaway Workshop, a more elaborate folding bench.
● And finally, Black & Decker's new Drop Leaf bench, smallest of the little benches.

The 29-inchers are successful examples of the compromises involved in designing a compact bench. While they are light and compact enough for easy storage and portability, they are also large enough to handle full-size jobs and heavy enough to have guts. You can take them to the job, set them up in seconds, and use them with confidence. They'll hold work of almost any shape, either between their jaws or between their swiveling plastic pegs. When the job is done, they fold up neatly and store away until next time.

Besides serving as both bench and vise, these tools also make good clamps. They're especially useful when edge-gluing two or more boards together.

Wall-mounted bench

Hirsh's Foldaway Workshop: This is not a *portable* bench: It's designed to be permanently mounted on your workshop or garage wall where its

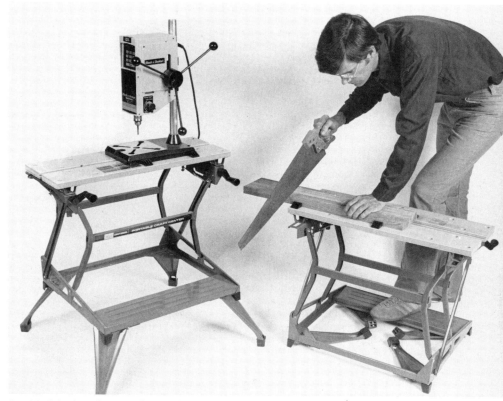

Dual-height benches can be used at normal bench height like the Craftsman on the left, or lowered to sawhorse height like the Workmate on the right. These identical benches sell for about $90, but may be discounted.

20 × 36-inch bench and integral woodworking vise can swing down for business. The tool's strong suit is security: the lockable cabinet offers plenty of space for tool storage on its perfboard door panel. The Foldaway Workshop can be installed with right- or left-hand door opening.

The little Drop Leaf Workmate is really too small to see much duty as a bench, and, strictly speaking, it's not really portable. Even so, I found it surprisingly useful. It really serves best as a vise, mounted at the end of a permanent workbench and level with its top. When you don't need it, it folds down out of the way. When you

do, it flips up and locks into position. Despite its small size, it has the same maximum clamping capacity as its larger brothers.

Like the larger tools, this one has horizontal and vertical V-grooves cut into its jaws so it can grip pipe, tubing, and other round stock. The Drop Leaf Workmate can also be used with most of the new accessories we'll be looking at next.

Bench accessories

Saw and miter guide. This attachment, plus the three others following, are offered by both Black & Decker and Sears, though there's no reason why they couldn't be used with the Wen and Hirsh benches. The saw and miter guide is an extruded aluminum track system that clamps between the jaws of your bench to guide your portable circular saw when making crosscuts and miters. When crosscutting it will take boards up to 17 inches wide, and when mitering at 30 degrees it will handle 7¾-inch stock. Maximum work-piece thickness is 2⅞ inches. To guide your miters the tool has a pivoting protractor arm, marked in 5 degree increments from 30 to 90 degrees.

The guide has two strong suits. First, since your saw is mounted in its normal attitude—blade down—this is a safe tool, unlike setups that mount your saw under a table with the blade exposed. Second, it greatly speeds your work, especially if you are making a series of multiple cuts. You just set the protractor arm, slide your work under the track, align your cut with the line-of-cut marker on your saw, and make your cut.

Accuracy? Good, but not perfect. The sliding track sections are a little loose, so the saw can wobble a bit from side to side. You can minimize this problem by pressing the saw toward the left edge of the track, just as if you were sliding it against a fence-type guide. The track also tends to flex up and down if you lean on the saw, and that can affect accuracy, too, until you develop the habit of using a light touch. Price for the guide is about $35.

Router and shaper guide. This is a clever router mount and work fence that—like the miter guide—clamps between the jaws of your bench. It will accept any router with a base 6 inches or less in diameter, and with three equally spaced mounting screws. Unlike those little shaper tables that take a router mounted bit up, this tool keeps the cutter bit completely covered. The price you pay for this safety factor is the inability to work stock any thicker than 1⅝ inches. Otherwise, the router guide is accurate, easy to adjust, and easy to use. It costs about $20.

Clamp attachments. These two fast-acting clamps fit into the peg holes in your bench top and can be used for clamping glue joints, or to hold stock that is too large to be clamped by the bench itself. (You'll need to drill ¾-inch diameter holes in the Hirsh bench to use them.) They can handle stock up to 3¾ inches thick, pivot 360 degrees in their mounts, and are especially useful with

All Bench's four-position legs adjust individually for variable bench height and angle. All Vise (included) can be used with chocks to hold up to 27½-inch work pieces. Price of the model 4500 bench: $80.

Hirsh's wall-mounted Foldaway Workshop has a drop-down bench that clamps work pieces with extension bars. Cabinet is 50 × 24 × 7 inches closed; $190. Optional 21-inch cabinet (shown), $52.

Drop Leaf Workmate, mounted at the end of a stationary bench, will let you hold stock as long as your bench. Drop Leaf can also mount on wall studs, folds down out of the way when not in use (inset). Tool's 16-inch vise jaws open to 5¼ inches.

the oversize table shown in the sketch. Price: about $15 a pair.

Drill guide. This simple rig will take just about any ¼- or ⅜-inch drill made. Clamp it in your bench and it will help guide your drill at any angle from zero to 90 degrees. It has a pair of built-in spirit levels to help you get accurate work at those two angles, plus an adjustable depth stop. Besides serving as a mini–drill press, it also makes a good drill mount when you're working with grinding and buffing wheels. Price is about $18.

DIY improvements

In addition to these accessories, there are four other ways you can improve the portable benches:
● The plastic swiveling clamp pegs are hard and can mar your work. And their faces are slick, so your work tends to pop loose now and then. You can solve both these problems by contact-cementing rubber pads cut from an inner tube to the peg faces.
● The portable benches are heavy and sturdy enough for most work, but they may skid around a bit if you're doing

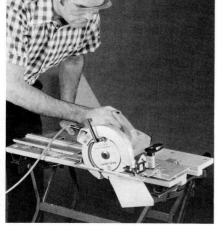

Saw guide has sliding track and pivoting protractor arm to help ensure fast, accurate cuts with your circular saw.

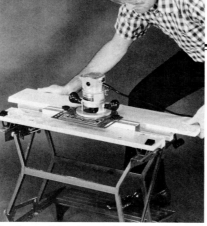

Router guide clamps between jaws of bench and turns your router into a shaper. Unit is sturdy, easy to use.

Swiveling plastic pegs give you a grip on odd-shaped work. Pad their faces with rubber to prevent marring.

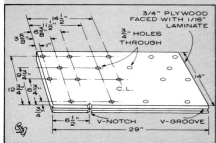

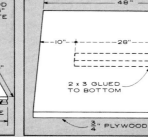

Bench accessories you can make: Extra-large rear jaw for portable benches (above right) increases table size, expands clamping capacity 6 inches to a maximum of 18 inches. Screw it in place of stock rear jaw. You can make one even larger than this if desired. Oversize table is easy to make. You mount it in place by clamping the 2 × 3 between bench jaws. Drill ¾-inch holes through table as needed for use with accessory clamps. Add extra weight to the bench as shown below for better stability when using this accessory.

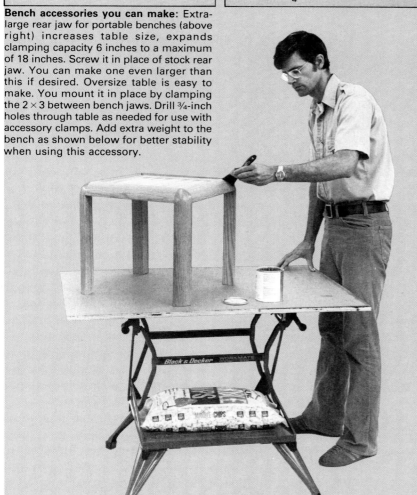

Drill guide, mounted in bench jaws, is used with accessory clamps, which hold work that won't fit in bench's vise.

heavy planing or the like. Solve this by slipping a plywood shelf into the frame. On this shelf you can place a 50-pound bag of sand or gravel (see photo).

● When you're using the bench as a wide clamp, the maximum capacity between swivel pegs is just 12 inches. That's not enough for gluing up wide stock. You can easily extend this capacity by making an oversize rear jaw, as shown in the drawing.

● The bench top is rather small (29 by 10 inches) for some jobs. You can make an accessory top like the one shown that clamps in place between the vise jaws. Note: Since this top overhangs the legs, it can be slightly unstable. The size shown is about as large as you can go and still maintain a stable bench. It's best used with a sandbag on the lower shelf for maximum stability—*A. J. Hand.*

MANUFACTURERS' ADDRESSES

Black & Decker, Dept. MM, 701 East Joppa Road, Towson, MD 21204; **The Hirsh Co.,** 8051 Central Park Avenue, Skokie, IL 60076; **Sears Roebuck and Co.,** Sears Tower, Chicago, IL 60684; **Wen Products, Inc.,** 5810 Northwest Highway, Chicago, IL 60631.

a pro's choices for fine woodworking

You don't need a roomful of tools to do fine woodworking. When I started building furniture, I owned a pad sander, scroll saw, router, and bench-mounted disc sander. I won't try to explain the rationale for that particular selection, but I built both Sheraton and French Provincial roll-top desks with it.

When I build a piece of furniture, whether for personal use or on commission, my interest is in getting the job done right and doing it in the minimum time. Don't get me wrong; I enjoy woodworking—I just can't afford to squander time doing it. Consequently, tools I've added to my collection over the years have been chosen because they contribute to doing the woodworking better, or faster, or both. That hasn't always meant power tools—or, at least, *big* power tools. Sometimes, I've found, it pays to think small. And sometimes, even in this day, the right hand tool will do a job better than any power tool you might use.

Among the most useful hand tools and small power tools I've added to my woodworking collection are some unusual ones that many of you may not have tried. I've selected a dozen of my favorites and demonstrate them on the following pages.

For me, these tools range from very useful to downright indispensable. Each can give you a pro's edge in your woodworking and can make a fine gift for woodworkers on your Christmas shopping list. Most are available at relatively modest cost. Good mail-order sources for all of them are listed at right —*Thomas Jones. Color photos by Greg Sharko.*

Sources for woodworking tools shown

Brookstone Co., 127 Vose Farm Road, Peterborough, NH 03458; **Constantine's** 2065 Eastchester Road, Bronx, NY 10461; **Craftsman Wood Service Co.**, 1735 W. Cortland Court, Addison, IL 60101; **Dremel**, 4915 21st Street, Racine, WI 53406; **The Fine Tool Shops, Inc.**, 20–28 Backus Avenue, Danbury, CT 06810; **The Foredom Electric Co.**, Bethel, CT 06801; **Garrett Wade Co.**, 161 Avenue of the Americas, New York, NY 10013; **Frog Tool Co., Ltd.**, 700 W. Jackson Boulevard, Chicago, IL 60606; **Leichtung**, 4944 Commerce Parkway, Cleveland, OH 44128; **E. C. Mitchell Co., Inc.**, P.O. Drawer 607, Middleton, MA 01949-0907; **Porter-Cable Corp.**, Jackson, TN 38301; **Sears Roebuck** (any catalog store); **Shopsmith, Inc.**, 750 Center Drive, Vandalia, OH 45377; **Woodcraft Supply Corp.**, 313 Montvale Ave., Woburn, MA 01888; **The Woodworkers' Store**, 21801 Industrial Blvd., Rogers, MN 55374.

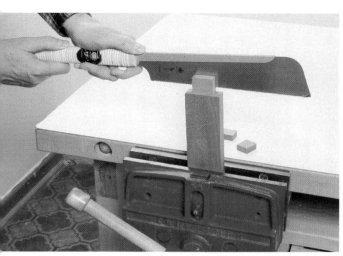

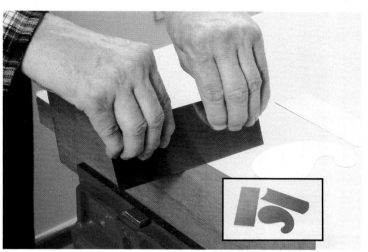

Dozuki saw, also called a tenon or dovetail saw, is efficient, easy to use, and accurate for light-duty work. Because it cuts on the pull stroke, the tensioned blade can be very thin—only 0.011 inch, compared with 0.03 inch for a typical backsaw. The thin blade and 0.005 inch of set produce a kerf just over 1/64 inch wide. When sawing, you put the blade *on* the pencil line, without worrying about which side is waste. The very narrow kerf also makes for easy sawing because so little wood is converted to sawdust. With 18 teeth per inch, the saw cuts very cleanly. The Dozuki is great for last-minute corrections, such as trimming 1/16 inch from a slightly long tenon. It's also a fast way of forming tenons, even if you have power tools, when you're making only one or two. Sharpening the very hard, closely spaced teeth on the laminated steel blade of a Dozuki saw is tedious, difficult, and requires a special file. Therefore, I like Garrett Wade's convenient replaceable-blade Dozuki saw. Cost: about $19; replacement blade: $9.

Hand scrapers: Once you get the hang of how to sharpen and use them, you'll wonder how you got along without one. A single scraper can be used for rough and finish work. Used after semi-finish sanding, a hand scraper can cut surface-preparation time in half. You can use a scraper for cleaning off dried glue and for shaving off even partially dry sags and bubbles in varnished surfaces. However, it should not be used for removing hardened paint; that would rapidly dull the edge. You can use a hand scraper to remove sander marks from a board edge without rounding the edge—an almost impossible task with a pad sander. But be advised: A new scraper is not ready to use. You must first file, stone, and hone the edge smooth, then put a burr on the edge with a burnisher. It's the burr, or hooked ridge, that does the work. Hand scrapers are sold, individually and in sets, by all tool mail-order houses. A typical three-piece set (see inset) will handle all surfaces except moldings and costs about $8.

Copydex Jointmaster (left) is a precision-made sawing jig that will let you make a wide variety of wood joints by hand with close-to-power-tool accuracy. A selector head and adjustable bracket automatically gauge the width of 90-degree joint recesses, with automatic compensation for the saw kerf. A handy wedge block provides an automatic 8-degree setup for making dovetail joints. Once it's set up, you can make multiple pieces without remeasuring. The Jointmaster is designed for a backsaw having minimal set. It's about $36 from Brookstone.

High-speed, small-displacement pad sanders (above) are a world away from conventional orbital/straight-line sanders. The motion of these two is orbital only, but the 12,000 orbit-per-minute speed and 1/10-inch pad displacement give swirl-free results. With fine paper you can get a near-mirror finish. You can move the sander with or across the grain. Porter-Cable model 330 Speed-Bloc is about $65; Makita's model BO4510, $70.

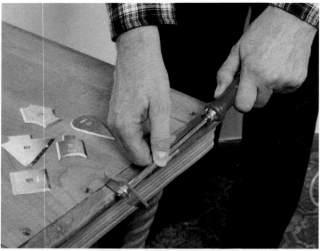

Molding scrapers are far more efficient than sandpaper for removing tool marks and other flaws from moldings. The eight blades in this set will handle most shapes. Blades are sharpened like chisel blades, not burred like hand scrapers. While molding scrapers are made primarily for preparing and refinishing moldings, you will find many uses for them in your shop. The Garrett Wade set shown costs about $28.

Dremel's 4-inch tilt-arbor table saw (model 580) is not a toy. Powerful it's not, but it is versatile, handy, and accurate. And it's so quiet I can run it late at night. If what I have to cut is not too big for the saw's 10 × 12-inch table and 1-inch capacity, I will use the Dremel instead of my 10-inch table saw. It's beautiful for rabbeting (a little slow with multiple passes for dadoing). I have built a tenoning jig for it and am planning other jigs. Two blades are available: a thirty-tooth combination and a 100-tooth fine blade. The saw is widely available for about $100.

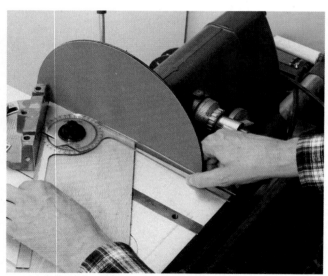

Mite-R-Gauge: Measuring and setting precise angles is important in many woodworking projects, and it can be difficult to accomplish. Interpolating inscribed settings on power tools can be hazardous to your accuracy. The usual method of using a sliding bevel with a protractor requires two steps and provides a fine opportunity to lose accuracy in the transferring process. Shopsmith's Mite-R-Gauge protractor/bevel ($16) solves the problem by letting you measure, lock, and read either inside or outside angles in a single step. The 10-inch arms take all the problems out of setting accurate table-saw and sander angles. **Maxi-Clamp:** Nobody's shop ever has enough clamps. And no matter how many clamps you have there will be some jobs you can't get hold of, and some repetitive jobs where you can go mad clamping and unclamping. Shopsmith's Maxi-Clamp will come to your rescue. If what you are trying to clamp has solid surfaces, you can, with ingenuity, put together the seventy parts of the Maxi-Clamp system—pressure feet, rods, fasteners, clamping jaws, and multihole junction blocks—in various combinations that will hold, push, pull, or position just about anything. For example, I recently used the system to assemble spline-mitered ogee feet for a cabinet. With the Maxi-Clamp I could hold the ends in exact position while applying pressure only to the mitered glue line. Another job was assembling fourteen drawers, all the same width and depth, for a tool cabinet. With that many, you want to glue each drawer in one shot. I did, with the setup shown. The Maxi-Clamp system costs about $50. After you get one, buy additional washers, hex nuts, and ⅜-inch threaded rod—a couple of six-foot lengths standing in the corner let you cut extra-long rods, and you can clamp almost any object that comes along. However, the Maxi-Clamp system is not a replacement for C-clamps, wood-jaw clamps, and pipe clamps.

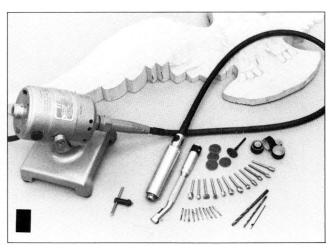

Foredom hand grinder (above): Hand grinders are useful for many operations—carving, drilling, sanding—particularly in inaccessible places. But Foredom's flexible-shaft models are the handiest of the lot. These miniature power tools are made in six motor styles with fifteen interchangeable handpieces. I have the No. 30 handpiece with key-type geared chuck, which will take twist drills and burrs, and the like, with shafts up to 5/32 inch. I also have the No. 55 contra-angle dental-type handpiece—very handy for getting into tight places with short latch-type mandrels and burrs. Foredom tools are carried by several mail-order tool houses. Bench-style and hang-up motors with electronic foot controls run approximately $200. The No. 30 handpiece costs $48; and the No. 55, $83. **Record doweling jig:** Most doweling jigs are far from versatile. Several will only locate dowel holes in the edges of boards or in framing; one will center dowel holes between edges only, unless you add a shim; another locates mating holes in boards for corner joints. Although my preferred method of drilling dowel holes is with a drill press or by means of horizontal boring with a Shopsmith, there are many occasions when I have to use a doweling jig instead. The Record model 148 doweling jig ($53 from Garrett Wade, $52 from Constantine) is the one jig I have not been able to stump with a doweling problem. It is the only one I know of that will locate dowel holes in the middle of a board. It will locate any holes needed to make casework furniture and will guide the drill for perpendicular drilling. As it comes, the English-made doweling jig will locate holes anywhere on any surface of a board up to 6 inches wide. Additional guide rods are available to increase width capacity to 12 and 18 inches. But you could drill holes across the center of a 48 × 96-inch plywood panel by substituting 3/8-inch drill rod for the guide rods. The tool comes with 1/4-inch, 3/8-inch, 6-mm, 8-mm, and 10-mm drill guides (5/16-inch is available).

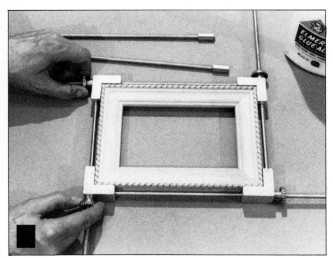

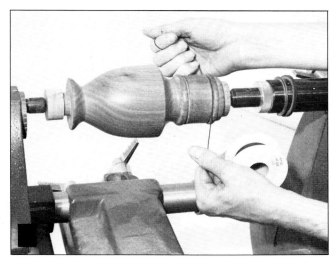

Miter-frame clamps (above) solve a standard woodworking problem: how to apply clamping pressure on a mitered joint for a picture frame or box. Corner clamps won't do it. They will position the parts, but they won't apply pressure. This miter-frame clamp will clamp all four corners in one shot. The aluminum-alloy corner blocks will not mar the work. Parts will be held with absolute accuracy. And if your miters are not exactly 45 degrees, or if one side is 1/16 inch short, you will know it. With additional threaded rods cut to length, you can extend the clamp to hold any-size frame or box you can lay flat on your bench or the floor. Some versions of this versatile clamp use 1/4-28 threaded rod, which makes it difficult and expensive to rig such extensions. I like the version that uses common 1/4-20 threaded rod. With it, economical clamp extensions can be made with the threaded rod available at hardware stores and readily obtainable hex coupler nuts. The miter-frame clamp's usefulness is not limited to mitered joints. Any kind of box or carcass joint can be clamped. Incidentally, with this clamp and accurate mitering you can put picture frames together with glue alone. Miter-frame clamps, with 18-inch rods, are available for about $18 from Constantine and others. **Abrasive cords** (above) have their biggest home-shop application in cleaning the residue of old paint and varnish out of the deep grooves in turnings after using paint stripper. I find them handy for sanding the grooves of new turnings, too, because they do the job a lot more efficiently than strips of sandpaper (which seem to tear apart just as you get them started working). The abrasive-impregnated cords come as round cords or flat tapes in 150 and 180 grits. They are made by E.C. Mitchell Co. and sold by the maker and many other mail-order sources for $8 to $10 per 25-yard spool. Smaller (30-foot) sample spools are available from the manufacturer—three spools for $10.

index